Patrick Moore's
Practical Astronomy series

Springer

London
Berlin
Heidelberg
New York
Barcelona
Hong Kong
Milan
Paris
Singapore
Tokyo

Other Titles in this Series

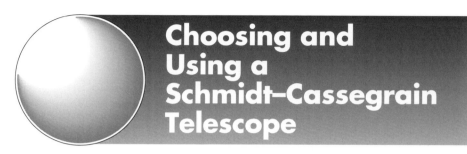

Choosing and Using a Schmidt–Cassegrain Telescope

A Guide to Commercial SCTs and Maksutovs

Rod Mollise

With 53 Figures

Springer

Cover photographs: By courtesy of Meade Instruments Corporation and
Celestron International

British Library Cataloguing in Publication Data
Mollise, Rod
    Choosing and using a Schmidt–Cassegrain telescope: a guide
    to commercial SCTs and Maksutovs. – (Patrick Moore's
    practical astronomy series)
    1. Schmidt telescopes – Handbooks, manuals, etc.
    2. Cassegrainian telescopes – Handbooks, manuals, etc.
    I. Title
    522.2
ISBN 1852336315

Library of Congress Cataloging-in-Publication Data
Mollise, Rod, 1953-
    Choosing and using a Schmidt–Cassegrain telescope: a guide to
commercial SCTs and Maksutovs/Rod Mollise.
        p. cm. – (Patrick Moore's practical astronomy series,
        ISSN 1431-9756)
    Includes index.
    ISBN 1-85233-631-5 (alk. paper)
    1. Reflecting telescopes. 2. Catadioptric systems.
    I. Practical astronomy.
QB88 .M65 2001
522'.2–dc21   00-061904

Patrick Moore's Practical Astronomy Series ISSN 1617-7185
ISBN 1-85233-631-5 Springer-Verlag London Berlin Heidelberg
a member of BertelsmannSpringer Science+Business Media GmbH
http://www.springer.co.uk

Typeset by EXPO Holdings, Malaysia
Printed and bound at the Cromwell Press, Trowbridge, Wiltshire
58/3830-5432 Printed on acid-free paper SPIN 10912140

Preface

The modern Schmidt–Cassegrain telescope is an incredibly capable tool for the amateur astronomer. But it is becoming a very complex one, with each new model boasting more features, more gadgets, and, increasingly, more reliance on computers. The manufacturers do their best to make their telescopes user-friendly and their manuals accessible, but nothing can substitute for straight talk and experienced tips from brother and sister amateurs. And that's what this book is all about. It's a compilation of the SCT wisdom I've picked up since the exciting evening in 1976 when I unpacked my first Orange-tube C8.

No work of this kind can possibly be the product of one person. I owe what I know about astronomy in general and SCTs in particular to the wonderful teachers I've had over the years: "teachers" meaning not just classroom instructors, but also those seasoned amateur astronomers who took a few moments to help a confused and anxious beginner sort things out.

There are many people I need to thank, but at the top of the list are the members of my Internet "sct-user" group, and especially the following friends who went out of their way to offer their encouragement and knowledge: Leonard Akers, Dave Bird, Alan Bland, Al Canarelli, Phil Chambers, Dan Cimbora, Bruce Cloutier, Steve Coe, Michael Covington, Michael Cunningham, Paul Goelz, Marcie Greer, Joe Hartley, Phillip Hosey, Simon Hastie, Mike Hosea, Walter Knapp, Michael McNeil, Peter Moreton, Joe Morris, Gary Otteson, Jeff Schaub, Larry Span, Glen Speck, David Tetreault, Greg Thompson, Rick Thurmond, Bob Van Deusen, and Russell Whigham.

My most heartfelt thanks, though, go to the two people without whom this book would honestly not exist. My best friend and long-time observing companion, Pat Rochford, spent quite a few hours of his precious spare time checking the manuscript, as did my wonderful wife Dorothy, the brightest star in this

astronomer's sky, and my constant inspiration. If you
enjoy what follows, they deserve the largest part of your
appreciation.

Rod Mollise
Selma Street
April 2000

Contents

Chapter 1

Why a CAT?

If you have even the slightest interest in the fascinating world of telescopes and astronomy, you have probably seen one of those short, stubby and wonderfully high-tech looking Schmidt–Cassegrain telescopes (SCTs). If, as a new amateur astronomer, you've been reading the astronomy magazines – *Astronomy Now, Sky and Telescope,* or *Astronomy* – you have no doubt been mightily impressed by the many full-page, full-color advertisements the SCT manufacturers run in every issue. Reading the enthusiastic ad copy, you'd think the Schmidt–Cassegrain is the only telescope worthy of your consideration. This may or may not be true, but putting aside advertising hyperbole, the SCT *may* well be the best and most versatile telescope for the average amateur astronomer. As we'll see, SCTs do have weaknesses as well as strengths, but the fact that these portable observatories have claimed a very large share of the telescope market over the last 25 years makes it obvious that there's *something* good going on here. Advertising copy alone wouldn't account for the astounding continuing popularity of these instruments.

For the novice, at first glance the SCT or CAT (short for *catadioptric*, a type of telescope containing lenses and mirrors) is both familiar and strange-looking. It appears to be a little like the telescope we're used to seeing in the movies and on TV. Its eyepiece is in the right place at the bottom of the tube, it is perched on a tripod, and it can be pointed at the sky. But there the similarity ends. The main impression given to the prospective SCT buyer is that this thing is an *absolute maze* of gears, dials, switches, and lights. Daunting.

Even frightening! But nothing could really be farther from the truth. The SCT is one of the most user-friendly and simple to operate telescopes on the market today. And capable? There is nothing out there that can match the SCT for versatility.

Why do amateur astronomers love the Schmidt–Cassegrain telescope? Simply because it is capable of doing many things well. Other telescope designs may have advantages for certain applications, but the key to the CAT's amazing popularity is its adaptability. An SCT is a "system." Numerous accessories and add-ons are available for these telescopes because Schmidt–Cassegrain manufacturers, unlike their counterparts in the computer industry, have willingly standardized many of the design features. For example, a camera adapter designed by Meade (one of the two big SCT manufacturers) will fit and work just as well on a telescope made by their chief competitor, Celestron. This makes things easy for the equipment purchaser and also encourages third party development of accessories and equipment for these telescopes. Some of the best add-ons for SCTs are made not by Meade or Celestron, but by members of a growing number of SCT accessory makers. If you can possibly imagine it – focusing motors, computers, camera adapters, tracking motors, and much, much more – it is probably available for your CAT. And you can be assured it will fit and operate as designed. This wealth of accessories also ensures that your telescope will grow with you in your new hobby. You may start out with nothing more in mind than getting a good look at the Moon. But, if amateur astronomy really "takes" with you, as it does with many, you may find yourself taking astrophotos a few years hence. And the wonderful thing about the SCT is that it is just as capable of doing complex tasks like photography as it is of giving you a pretty image of that good, old Moon!

SCT Capabilities

"Jack of all trades, master of none." An old aphorism, and sometimes true. But *not* true in the case of the SCT. While nothing in the design of the SCT is astoundingly innovative, the basic layout of this scope is extremely sound and allows these instruments to achieve their renowned versatility. The SCT doesn't just do a *lot* of things – it does a lot of things *very well*. Here are some of the things *you* can do with *your* CAT:

General Visual Observing

Whether your interest is viewing the Moon and planets or cruising the dark depths of space in search of faint fuzzies – star clusters, galaxies and nebulae – the CAT is very well suited for the visual observer. Part of the reason for this is its generous aperture, the large size of its mirror in relative terms. Quality is important in telescopes, but the sort of image you get in your telescope, whether of a planet or of an unimaginably distant galaxy, is more than anything else dependent on the size of your telescope's main lens or mirror. Small refractors, the traditional lens-type telescopes, can often provide beautiful views. But for most tasks your SCT will beat refractors every time, even very sophisticated refractors costing many times more than your humble SCT. Why? Because even the largest refractors are usually limited to an aperture size of 6 inches (150 millimeters). The average SCT, on the other hand, features an aperture of 8 inches (200 millimeters). A 6 inch refractor can provide incredibly wonderful images. But most of the time you'll see more detail with your 8 inch SCT! This is not surprising. An 8 inch telescope delivers almost twice as much light to your eye as a 6 inch.

And it's not just the optics which make an SCT a wonderful tool for the visual observer. Almost all Schmidt–Cassegrains are equipped with clock drives. These are motors that allow the telescope to track the stars. A big reflecting telescope, a Dobsonian, with its enormous mirror can provide wonderful views. But the SCT owner with her 8 inch is often able to detect more details than the Dob owner. Why? Because the image can be viewed in comfort. The scope keeps that planet or star cluster centered in the eyepiece and all you have to do is look. No nudging the telescope – and possibly losing that much-searched-for galaxy in the process.

Photography

One glance at the beautiful color pictures in the astronomy magazines and a novice astronomer wants to start taking astrophotos. Astrophotography is a difficult, maddening and rewarding pursuit. Just about any scope can be adapted to take photos of the deep sky or the Moon or the planets. But most telescopes will require fairly extensive modifications before they're really up to the task. Refractors must be provided with guide telescopes and often need better mountings and clock drives. Newtonian reflectors may require that their mirrors be repositioned in the telescope tube before a camera can even be brought into focus. In the past, making a telescope ready for photography often meant the amateur had to rebuild it from the ground up. But not the SCT. These telescopes were designed to be able to take sky pictures easily. With the addition of a few simple accessories, all fork-mounted SCTs can be used without modification for astrophotography. Take a mid-range, or even a bargain SCT model; add a 35 millimeter camera body, an off-axis guider (a special type of camera adapter), and a few other items, and even a relative novice can, almost from the beginning, take photos which will compare quite favorably with the pictures the astronomy magazines print.

This is not theorizing, nor is it manufacturers' rhetoric. I proved this to *myself*. I hadn't tried astrophotography in 20 years; I had found the process just too frustrating. I'd spent weeks fine tuning my old Cave Newtonian reflector, and suffered many hours of shivering Ozark Mountain cold. And for what? A few blurred black and white pictures of the Orion Nebula (I knew it was the Orion Nebula, but none of my friends could tell from my photos). Even after buying my first SCT I never returned to photography. Just too much work for *no* results. But a few years ago I bought my current CAT, a Celestron Ultima C8, a telescope designed from top to bottom with the photographer in mind. Seemed like a shame not to put a camera on it. OK, I'd give it a try.

I purchased a very inexpensive off-axis guider, set the telescope up in a friend's light-polluted backyard, pointed the scope at Orion, squinted through the guiding eyepiece and exposed away. Seemed pretty easy. Much easier than taking pictures with my old Newtonian. In fact, it really seemed *too* easy. I just wasn't convinced that fumble fingered me could take deep space photos. The next day I made a quick trip to the photo finisher and handed over my roll to the lab technician. I waited in agony for the mini-lab to spit out my pictures – I didn't expect much, but I just couldn't help *hoping.* When the technician brought my prints over I could hardly believe it! There was the Great Nebula, big and beautiful. The wondrous colors, brilliant reds and blues, were simply unbelievable. No, it *wasn't* perfect, but it was pretty darned good. I could hardly wait to share my images with my astronomy friends and my wife. Let me emphasize that this doesn't say much about my still lacking skills as an astro-photographer. What it says *worlds* about is the ease of astronomical picture taking offered by the SCT.

Advanced Applications

But the answers to the "what is it good for?" of a CAT don't stop with visual observing and astrophotography by any means. Since these scopes *are* so versatile, they can be used in some pretty serious scientific endeavors, and *are* being used for these tasks by amateur and professional astronomers every day. Want to hunt for asteroids? SCTs operated by amateurs using sensitive electronic CCD cameras are finding new ones every single night. Interested in double stars? An SCT, with its generous focal length, makes measuring these stellar systems' separations and position angles very easy. Variable stars? Hang a CCD camera or a photometer off the back of your CAT and have a field day checking the brightness of the distant stars. Want to do these things, but don't like going out into the cold dark? There are several modern SCTs that can be remote controlled from the warmth and comfort of your computer room.

SCT Liabilities

I'm enthusiastic about my CATs, and I don't mind saying so. But after using these scopes for 25 years and having owned a number of them, I can admit that they're not perfect. What is? The SCT design, like the design of any telescope, is a compromise. I don't think these minuses do much to reduce the telescopes' overall strengths, but you should be aware of them. Expect to be *made* aware of them at any gathering of amateurs. There is a faction within the amateur community who seem to delight in SCT-bashing. There may be some truth to some of these claims, but on closer examination, most prove to be based more upon supposition than actual telescope use in the field.

Contrast Problems

A Schmidt–Cassegrain is an *obstructed* telescope. This means that it has a secondary mirror – an obstruction – placed in its light path. According to optical theory, this may cause degradation in contrast, something which is particularly troubling for the planetary observer. Most reflecting type telescopes do have this central obstruction due to the placement of their secondary mirrors. But some critics will tell you that the SCT is the prime offender in this area because its secondary mirror, due to the basic design of the telescope, must be relatively large, often comprising 30% of the total aperture of the telescope.

The simple fact of the matter is that *any* obstruction placed in the light path, no matter how small, will damage contrast to some extent. Even a custom Newtonian reflector with a very small secondary mirror has lost out compared to the unobstructed refractor. The question is, "Does the larger secondary mirror of the SCT make things much worse?" Based upon my 25 years of experience the answer is: "no," or at least "not much." Listen to some of the "experts," and you'll start believing that because of this supposed problem the average SCT must deliver an image about as sharp as that offered by a 60 millimeter department store refractor. But then go look through an SCT and prepare to be amazed. Despite talk about a large secondary robbing light, sharpness and contrast, the job these telescopes can do on the planets is simply amazing. I've had refractor fans stare in amazement at the image of Saturn offered by my Ultima 8 SCT. An image that is sharp, contrasty and beautifully detailed even at 500x.

Collimation

A Schmidt–Cassegrain telescope can only perform well if it is *collimated*, that is, if its two mirrors are properly aligned. SCTs are particularly sensitive to misalignment of their mirrors. The slightest maladjustment of the SCT optical train can completely destroy the quality of planetary images. The good news is that SCT alignment is easy, and once collimated, an SCT may hold this alignment for many months. My personal SCT, for example, was still in perfect adjustment after traveling hundreds of miles over pothole-laden US interstate highways in the back of a rental truck on the way to the Texas Star Party.

Small Aperture

If you're a novice amateur, the 8 inch mirror in the average SCT may seem big, especially if you've been using a 2 or 3 inch refractor previously. But there is no denying that in the amateur astronomy world of today an 8 inch mirror is on the small side. This will hit home at your first big star party when somebody pulls up next to you and sets up his 30 inch Dobsonian telescope. There is no way that your 8 inch can compete with a 30 inch on the deep sky. But an 8 inch aperture telescope is plenty big enough to keep you going and interested for a lifetime. It will show you literally thousands of deep sky objects – probably more than you'll ever get around to observing. Many of the brighter objects like those that comprise the Messier catalog will show considerable detail, too. Galaxy M51 will show off its spiral arms, the Great Orion Nebula will seem to flow across the sky and into infinity, and Globular Cluster M13 will easily break down into countless diamond-dust stars. Remember, too, your CAT is capable of doing many things. That big Dob is really only suitable for visual observing.

Are SCTs *really* Portable?

I used to think they were. They are certainly a lot easier to load, transport, set up and break down than my 60s–70s style Newtonian reflectors were. SCTs are promoted as being very portable by their manufacturers, but I would change this to "very transportable." Except for the lightest bargain-model CATs, most of these scopes are not exactly lightweights, with a weight of 50 pounds (20 kilograms) for the combination of telescope and tripod being common. They also require quite a bit of setup compared to some other scope types. In my experience, the usual star partystar party routine goes something like this:

Drive up. Unload tripod and place it so that it is oriented to the north. Dig out and use a compass if it's too light to see Polaris, the North Star. Unload the scope in its case from the trunk. Next to me the owner of a 12 inch aperture Dobsonian has placed his rocker box (mount/telescope base) on the ground, gotten the tube out of the back seat and placed it in the rocker. He's ready to go. I grit my teeth. Opening the case, I retrieve the wedge attachment bolts and insert one into the scope base. I lift my heavy Ultima SCT onto its wedge and adjust it so that I can insert two more wedge bolts. I tighten these down. I find the star diagonal and visual back and mount them on the scope rear cell. Next I search through my accessory box for my dew shield, and, if necessary, my dew heaters. I attach these and start setting up my external battery power supply if I need one on this night. As it's now dark, I start the polar alignment process. My neighbor has already viewed several objects. After a few hours, the sky clouds over. Time to pack it in. I reverse the setup process. As I'm loosening the first wedge bolt I look over at the Dob owner. His scope is packed away and he's drinking a cup of coffee while grinning at my plight!

All of this is quite true. But it is also true that when you do have your SCT set up, you have at your disposal a virtual portable observatory, ready to take on any task. A CAT does take some time to assemble, but it can be packed easily into even the smallest vehicle. Unlike the Big Dob owner, your choice of telescope doesn't necessarily dictate your choice of vehicle.

Is a CAT for Me?

A Schmidt–Cassegrain telescope is a wonderfully capable telescope. But is it the *right* telescope? For *you*? Only you can answer this question, but the following should help.

The SCT May Be Your Scope If

You haven't specialized in a particular area of astronomy, and probably don't intend to. An astro dilettante like the author will go from visual planetary observing today, to galaxy hunting tomorrow and scientific data taking next week. You need a scope you won't outgrow, one that can change as your interests change.

You want to take pictures. But you don't want to spend a lot of money. Tens of thousands of dollars can easily be spent in the quest of the perfect astrophotography setup. A relatively inexpensive SCT can take marvelous photos of the sky, and is probably capable of better photographic performance than most astrophotographers will ever need.

You want to be able to do a little observing and photography from your light-polluted backyard and travel to dark sites for real picture taking and deep sky observing. The virtue of the SCT here is that while it performs amazingly well from dark sites and is transportable enough to make trips pleasurable affairs, it can also do surprisingly well from the backyard. Its long focal length is a big help in taking pictures from the average suburb (a short focal length wide-field telescope may have its film fogged from sky glow before much of an exposure can be accumulated). Long focal length can also be a big help for that staple of the urban observer, planet watching. The f/10 optical system means that high magnifications appropriate for producing large image scales on the planets can be achieved without Barlow lenses or uncomfortably short focal length eyepieces.

You're a "techno geek." You like gadgets and electronics and wouldn't dream of owning a noncomputerized telescope. If computers and electronics are your thing, you've come to the right place! There are SCTs that guide you to the object you're interested in, SCTs that move themselves from object to object and even SCTs that do the work themselves, taking images of selected objects while you doze in a warm bed miles away.

You're physically challenged. Even a 6 inch Dobsonian telescope is impossible for you to handle. SCTs win hands down here. Their short tubes make them naturals for use by physically challenged, slightly built or young people. An 8 inch SCT may be too much for some of these folks, but SCTs are also available in small, ultra portable sizes: 5 inch, 4 inch and 3.5 inch SCTs and similar scopes are a staple of the market. These small CATs still offer that portable observatory experience with exquisite optics, built-in drives, computers and an array of accessories similar to that offered for their big sisters (often SCT accessories used on the larger scopes will work on the smaller ones as well).

An SCT May Not Be For You If

You just want to look, and mainly at deep sky objects. You don't care about taking pictures – you'll buy a book or view them on the Internet. You want to see galaxies and nebulae and you want to see them live, looking as much as possible like photographs. Think "20 inch Dobsonian" if this describes you, not "SCT."

You're an advanced photographer or CCD imager, and you're particularly interested in wide-field shots. You want perfection – and you have the money to pay for it! You could still be happy with a top-of-the-line SCT, but you may be happier with a big apochromatic (color free) refractor.

You don't like gadgets of any kind, and they don't like you. The thought of hauling a battery pack, much less a computer, into a damp field gives you pause. Your motto is "the simpler, the better." Look for a medium-sized (6–10 inch) Newtonian telescope, particularly a Dobsonian. It'll keep your blood pressure down and your hairline intact.

Still can't decide? I urge prospective SCT owners to actually try one of these telescopes before getting their hearts set on one. Most cities and towns of any size in the U.S. and Europe have active astronomy clubs. Members of most societies will be only too happy to help you with your decision. Quite a few amateurs will consider it their *personal mission* to help you select the right scope (that's the nature of the people in this hobby). You'll no doubt run into quite a few SCT users who'll be only too happy to let you be "copilot" on their scope at the next star party.

No club? No SCT owners you can find locally? I urge you then to at least try to seek out a shop selling these telescopes; even if you have to travel to another city. It is very important that you at least get an idea of how big these scopes are (again, *much* bigger than they look in the advertisements). It may be that you need a smaller SCT. Or possibly you'll find that you can easily handle a 10 or 11 inch or larger CAT. In any case, you do have this book. If you stick with it, by the end you'll be as competent as can be without hands on experience, in not only choosing, but setting up, using and enjoying a Schmidt–Cassegrain telescope.

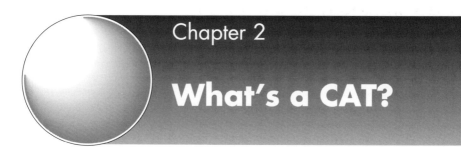

Chapter 2

What's a CAT?

What makes an SCT work? Optics are the heart of any telescope. To a very great extent everything else about an instrument, its mounting, its capabilities, even its price, is determined by the telescope's optical design. So, let's begin our examination of the inner workings of these cosmic voyagers with a look at telescope optics. The Schmidt–Cassegrain is an optical hybrid. It combines some of the best characteristics of two different types of telescope, the refractor and the reflector. The secret to understanding the optics of our CATs is an understanding of the way these two simple telescope designs work.

In the Beginning

The simplest telescope of all is the refractor, the lens-type telescope. It was a refractor that Galileo Galilei first turned to the heavens on that wonderful Italian evening in 1609. Galileo didn't invent the telescope, and may not even have been the first person to use one for astronomy, but he brought the idea of using a telescope to observe the Moon, planets and stars to the attention of the Renaissance's growing scientific community. We can only wonder what took man so long to come up with this idea. The basic telescope is, after all, a very simple affair.

The secret of the telescope is found at the end of the refractor's tube. In this design a large lens is used to gather light (see Figure 2.1). It may be made from a

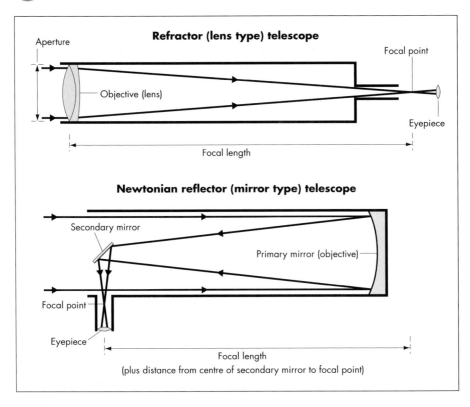

Figure 2.1. Basic refracting and reflecting telescope.

single piece of glass, or it may be composed of several *elements*, but its only purpose is to collect light, much more light than the tiny lens of the human eye. This light is formed into an image a distance from this *objective* lens. At this time the image of the heavenly body in the field of view of the telescope is still small, though it is now very bright. A magnifying lens is placed just beyond the telescope's focus point to enlarge the image and reveal fine details. This lens, the *eyepiece*, works like a household magnifying glass to enlarge the small image delivered by the objective lens.

At the most basic level, that's all there is to a telescope: a lens, the objective, to gather lots of light, and a magnifying glass, the eyepiece, to make this image bigger. Astronomers in the seventeenth and eighteenth centuries used this simple refractor to open up the cosmos to human view for the first time. Unfortunately, the Galilean refractors, and even the improved models that came after Galileo's day, suffered from some severe problems. The most devastating of these is *chromatic aberration*. A simple lens does not bring all the rays of light to focus at exactly the same point. This is a result of the refraction caused by light passing through the glass. Eventually a refractor would be made with much reduced chromatic aberration, but until that happened, refractor-using astronomers had to put up with an image that was blurred and ringed with spurious color.

Then a genius, perhaps the greatest genius the human race has yet produced, Isaac Newton, turned his mind to the problem and came up with a solution. Why use a lens? A lens isn't the *only* thing that can collect light. Why wouldn't a concave mirror do as well? Isaac Newton's telescope design replaces the refractor's convex lens with a concave mirror (see Figure 2.1). Light from celestial objects is collected by this mirror and reflected up the tube to a small flat *secondary mirror* that diverts the image out the side of the tube for easy viewing. Here, the image can be magnified by an eyepiece just as in the refractor. Using a mirror in this fashion completely eliminates the color problem.

The refractor and the Newtonian reflector seem quite different, but some of their characteristics are measured in the same ways. The diameter of a telescope's main lens or mirror is its *aperture*. The exact spot where the rays of light from the objective come to focus is its *focal point*. The distance from the lens or mirror to the focal point is the telescope's *focal length*. The ratio of the telescope's focal length to the diameter of its objective lens or mirror is its *focal ratio*. A 6 inch diameter mirror or lens with a focal length of 48 inches has a focal ratio of 8 (48/6). This is usually written as f/8 (or whatever the ratio is).

Birth of the CAT

We've been talking about Newtonians and reflectors synonymously. But not *all* reflecting telescopes, even the very earliest ones, have followed Isaac Newton's original and simple design. Throughout the history of these telescopes, opticians, both amateur and professional, have been experimenting with a wide variety of designs. Often the only thing two types of reflecting telescope have in common is that both use mirrors rather than lenses as their light-gathering elements. Two of these alternate designs, the Cassegrain and the Schmidt camera are the direct ancestors of our SCTs.

The Cassegrain

A Frenchman, Guillaume "Jacques" Cassegrain, came up with his innovative design for a telescope in 1672, only a few years after Newton demonstrated his first little mirror telescope. There is, however, no evidence that Cassegrain ever got around to actually building one of these clever instruments. Development of the Cassegrain telescope (now often known as the classical Cassegrain) really had to wait until mirror grinding and figuring knowledge had matured a bit.

Jacques Cassegrain's idea was this: take a concave mirror, as in the Newtonian, but figure it to a relatively short focal length. Drill a hole in this primary mirror. Place a small convex secondary mirror just inside the focus of the primary. Unlike the secondary mirror in the Newtonians, this mirror is not tilted at 45 degrees. Instead, it is arranged so as to direct the light rays back through the hole you've made in the primary mirror (see Figure 2.2).

This simple-seeming arrangement has a number of advantages over the standard reflecting telescope. The image is directed out the back of the scope, as in a refractor and viewing is very comfortable. There is no awkward observing position at the top end of a long tube as in Sir Isaac's telescopes. The convex secondary mirror also adds another level of convenience to the Cassegrain. The fact that it is convex means that it magnifies the image coming from the short focal length primary mirror. This causes the effective focal length of the scope to be multiplied. When this magnification is combined with the folding-back of the light path, the result is a long focal length instrument that yields high magnifications and reduces optical aberrations in a

Figure 2.2. The classical Cassegrain reflector.

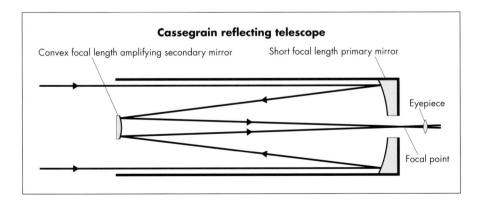

Cassegrain reflecting telescope

Convex focal length amplifying secondary mirror　　　Short focal length primary mirror

Eyepiece

Focal point

tube much shorter than would normally be required. The tube of a 6 inch Newtonian with a focal ratio of f/10 will be around 60 inches long. But a 6 inch f/10 Cassegrain may have a tube length of half this, or even less, depending on the exact curve and magnification factor of the secondary. With a short tube, the problems involved in designing and building a solid, yet light, mounting for the telescope are greatly reduced.

Sounds good. Is the classical Cassegrain the perfect telescope? Hardly. Along with these pluses come some real minuses. The Cassegrain suffers from fairly severe field curvature. That is, the edge of the field is at a different focus from the center. When the central area of the image is in focus, the edge of the field is blurry. Astigmatism, another optical problem inherent in these scopes, can reduce sharpness across the entire field. Cassegrains can also be very difficult to collimate. All of these problems can be alleviated if not completely eliminated. But because of these limitations, only rarely is a true classical Cassegrain used by modern professional or amateur astronomers.

The Schmidt Camera

In 1930, German optician Bernhard Schmidt conceived of a special project. Discussions with Mount Wilson Observatory astronomer Walter Baade led him to wonder about the feasibility of designing a wide-field astronomical camera. At this time it was becoming obvious that something of this nature was badly needed by astronomers. Telescope mirrors were getting bigger and bigger, and their focal lengths were getting longer. Longer focal lengths mean higher magnification and smaller fields of view. What the astronomers needed was a special camera-telescope with a very wide field to act as a *scout*, surveying larger areas of the sky. When something of interest was detected, the giant narrow-field reflectors could then examine it in detail.

Schmidt's design for this astrocamera sounds simple, though in actual practice it is not easy to implement. He began with an easy-to-produce spherical mirror. He ground a deep curve into this mirror to make the focal length short. But spherical mirrors at short focal lengths suffer from a very severe problem called *spherical aberration*. This is caused by the fact that light rays reflected from a sphere do not all reach focus at the same point. It is similar to the refractor's chromatic aberration and is what initially plagued the Hubble Space Telescope. It was obvious to Schmidt that he couldn't stop with a spherical mirror. He had to correct the aberration somehow.

What he came up with is the heart of the Schmidt camera, a special lens, a *corrector plate*, which was placed in front of the mirror. The shape of this thin glass lens causes the incoming rays of light to be bent just slightly, so that when they are reflected by the spherical mirror they reach focus together. The corrector adds the opposite amount of spherical aberration to the system, which causes the error to be canceled out. The focus of the Schmidt camera lies midway between the corrector plate and the mirror, so it would be a little difficult to place an eyepiece in the right position to view the produced image. But this didn't trouble Schmidt. He didn't envisage anybody looking through this telescope at all – it was to be a giant camera.

Putting It All Together – The Schmidt–Cassegrain

We have Cassegrains, and the first CATs, Schmidt cameras. How are these two different designs combined into the one used by our beloved SCTs? It had been obvious for a long time that the Cassegrain offered substantial advantages for the amateur astronomer and particularly for the astrophotographer. The short tube was the key. An 8 inch or 10 inch Cassegrain is exceptionally transportable; quite easily accommodated by even a light mount and tripod. Unfortunately, the Cassegrain's strong points were always overshadowed by its weaknesses. In addition to the problems with aberrations and alignment, a big strike against the Cassegrain is the difficulty involved in producing its mirrors, which require difficult-to-grind parabolic curves.

By the 1950s, a number of advanced amateur telescope makers had one eye on the Cassegrain, and one eye on Bernhard Schmidt's camera. The main thing holding the Cassegrain back was its hard-to-make, short focal length (often f/2) parabolic-shaped primary mirror. The Schmidt camera, in contrast, boasts an

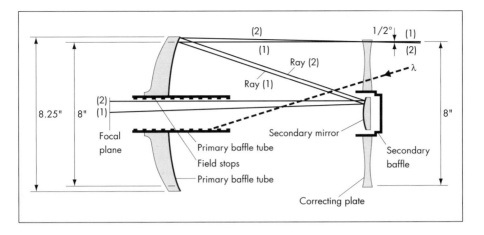

Figure 2.3. The
Schmidt–Cassegrain
(illustration courtesy of
Meade Instruments
Corporation).

easy-to-fabricate spherical mirror. What telescope
makers needed to do was add a *corrector plate* to
the Cassegrain. The result is shown in Figure 2.3. The
Cassegrain's parabolic mirror has been replaced by a
simple spheroid. The secondary mirror, which can
also be a sphere in this design, is now mounted in a
holder that is attached to the all-important corrector
plate. The corrected light rays coming from the
spherical mirror are reflected to the Cassegrain-like
convex secondary. The image is brought to focus
through a central hole in the primary, just as in the
classical Cassegrain. The final result is a telescope
with easy-to-make spherical mirrors, one which
combines long focal length in a short, easy-to-mount
tube. What could be simpler?

A lot of things, it turns out. When enterprising
optical workers decided to tackle the SCT, their first
problem was the Schmidt corrector plate. This very
thin, very mildly curved, and very difficult-to-grind lens
was not and is *still* not something that amateur
telescope makers, even fairly advanced ones, have
cared to try their hands at. Bernhard Schmidt *had*
developed a rather clever scheme to make the production of this vital telescope component a little easier,
though. Rather than trying to grind the complex
corrector curve by hand, Schmidt found that he could
"cheat" by placing the thin glass blank in a special
apparatus in which the circular glass blank forms one
side of a vacuum chamber. Just the right amount of
vacuum is applied to this chamber, causing the
corrector blank to deform, to bend inwards, by just
the right amount. From here, grinding the blank is easy;

the outward facing side of the corrector blank is simply ground and polished flat. When the vacuum is released, the corrector blank springs back, and, almost magically, the right curve has been ground into it. It *sounds* easy, but it definitely is not. In practice, many things can and do go wrong. But talented opticians persevered, produced SCTs with the right curves on their optical surfaces, and turned them on the heavens. And were badly disappointed.

There was yet another problem to be solved in the production of the SCT. It was soon obvious that the mirror part of the CAT couldn't be quite as simple as it first appeared. The two spherical mirrors were easy enough to produce, but the use of spheres results in very poor off-axis performance – stars at the edge of the field looked like blobs rather than points.

But the SCT was such a good idea that telescope makers kept trying. Opticians found there *was* a way to make the SCT field flatter. In fact, they found a number of solutions. The easiest to implement was making a secondary mirror with a curve that departs just a bit from a sphere. If a secondary mirror is ground with a curve that is almost, but not quite, a parabola, things get a lot better. This *aspheric* secondary mirror turns an almost-ran telescope into a powerful performer. Although the field of the f/10 SCT is not quite as flat as that of an equivalent focal length Newtonian, it is, with the addition of the aspheric secondary to the design, very sharp.

So the SCT had really arrived? Well, no. As if the difficult-to-make corrector plate wasn't enough, the poor amateur telescope maker now also has the task of making a complex, nonspherical secondary mirror. Most amateur astronomers threw their hands up, forgot about the CAT and went back to making and using good old Newtonian mirrors. SCTs were complex curiosities and might have remained that way except for the vision of one man.

Chapter 3

Inside a CAT

The Commercial SCT

Tom Johnson was a man with a mission: the resurrection of the Schmidt–Cassegrain. By the 1960s quite a few folks had begun to think that the SCT could be the perfect telescope for amateurs. Light, compact and with generous aperture, this CAT would be great – if only a way could be found to get it into the hands of users. Johnson, a Californian, was an expert telescope maker, and showed up at various amateur events early in the decade with a beautiful (and huge) 19 inch SCT. This wasn't that unusual. Telescope makers had been showing off their telescope *magnum opuses* at star parties for a while. But with Johnson, it was different. He had a plan. Using some new and innovative methods of manufacture, he dreamed of putting the SCT into *mass production*.

The corrector, it was obvious, would be the big snag in producing SCTs on a large scale, but Johnson had a method that simplified the process. The exact technique is still proprietary information, but the general procedure is common knowledge. Rather than placing a very precisely controlled and hard to regulate vacuum on the blank, Johnson's procedure calls for a vacuum to be called on merely to pull the blank against a very precisely made metal mold or master tool. This tool is in the exact opposite, the *inverse*, of the figure needed for a corrector plate. The glass blank is then ground and polished flat, just as in Schmidt's original procedure. When the vacuum is released, you have a perfect corrector plate. Obviously, the metal masters are difficult and expensive to make and must be very precisely done, so this is not something that would benefit an amateur telescope maker. But it made the establishment of an SCT factory possible.

That's exactly what Johnson did. He renamed his small electronics manufacturing business "Celestron Pacific" and by the mid-1960s this new company was offering a full line of Schmidt–Cassegrains for sale. These telescopes were not really aimed at the amateur, though. Celestron began by focusing on instruments for small colleges and institutions, like museums and planetariums, who had need of a good professional-level telescope but couldn't afford a large custom scope. This was a niche that Celestron was only too willing and able to fill with its new Schmidt–Cassegrains.

What were these new telescopes like? Mechanically and optically, much like the SCTs of today. There are differences, of course. The most obvious one is size. Celestron's telescopes ranged from a 10 inch all the way up to a massive 22 inch monster. Since these telescopes were generally intended for permanent installation in an observatory, the mountings, even on the smaller scopes, were more massive than the average CAT of today. They were furnished with large metal piers rather than tripods. These were beautiful telescopes, which featured gleaming white or light-blue aluminum tubes (hence the occasional reference today to a *white-tube* Celestron). I can remember admiring the advertisements Celestron ran in *Sky and Telescope* magazine in the 1960s and dreaming about owning one of these beauties. Realistically, I knew that they were far beyond my reach. You'd have to be a pretty wealthy amateur in 1967 to afford the pre-inflation US$2,500 the C10 commanded.

It didn't take long for this to change. In 1970 Celestron Pacific (soon to be known simply as Celestron or, more recently, Celestron International) began offering an 8 inch Schmidt–Cassegrain, the legendary original C8, at a price of about $1,000. This was a lot of money when you consider that most amateurs were still using inexpensive homemade Newtonians. But times were changing and more and more amateur astronomers were ready to pay a considerable sum for a telescope that really met their needs. The compact SCT was perfect for the modern amateur, who was more and more likely to have to pack up the telescope and travel far out of town to escape city lights. It also fed the growing amateur hunger for an astrophotography-capable scope.

Celestron has continued producing fine telescopes to this day, with the company emphasis now being squarely on amateur equipment. Over the years, Celestron's line has expanded to include a 14 inch SCT, an 11 inch and a 5 inch. But the mainstay remains the telescope that started it all, the C8 (now available in several models). Celestron has had a history of success, and shows every indication of continuing, but even if the company disappeared tomorrow, it and its founder, Tom Johnson, have earned at least some mention in the astronomical history books for finally bringing the amateur astronomer a modern, high-quality, affordable telescope.

So it went for a couple of years. Celestron was the only game in town if you wanted an SCT. There was no competition until 1971 when the Criterion Manufacturing Company, which had been producing Newtonian reflecting telescopes in the US since the 1950s, decided it was time to emulate Celestron. The company debuted two SCTs, an 8 inch and a 6 inch. In design, these scopes were very similar to the Celestron telescopes. But from the beginning, the Dynamax, as Criterion called its telescope, was plagued with serious optical and mechanical problems and was not able to garner the popularity that the C8 had enjoyed almost from its birth.

Then a company calling itself Meade appeared on the scene. The rise of Meade Instruments is one of those good, old-fashioned success stories Americans love so much. Meade began as John Diebel's one-person home-based business in the early 1970s, selling small imported astronomy accessories like eyepieces and focusers through tiny ads in the astronomy magazines. Meade didn't exactly take the world of amateur astronomy by storm, but Diebel kept at it, adding to and broadening his company's line of products and advertising steadily. After a couple of years this steady plodding started to pay off. Amateurs noticed that Meade was offering some pretty good eyepieces for bargain prices, something that was rare in the astronomy marketplace of the early 1970s. Meade's prospects advanced even further in 1978 when it finally started selling telescopes – 6 and 8 inch Newtonian reflectors at first.

Figure 3.1. The Meade SCTs (photo courtesy of Meade Instruments Corporation).

It was clear that Meade was a company with a lot of potential. But it was also clear that the market for the accessories, Newtonian telescopes, and old-fashioned achromatic refractors (another big product for the young company) was strictly limited. One thing appeared certain: the SCT was the wave of the future for small-telescope users, and the only way to really get ahead was to take on Celestron by producing a CAT. Meade, it was decided, would give it a try, even though that other popular Newtonian maker, Criterion, was currently in the process of failing the same challenge. After two years of development, Meade introduced its first SCT, the Meade '2080' in 1980.

The 2080 was hardly another Dynamax. The design was very similar to that of the C8, and in quality the telescope was much closer to Celestron's instruments than it was to the doomed Criterion scope. In some regards, the Meade even seemed *superior* to Celestron's famous original CAT. Meade had done its homework, taking two years to design a telescope and a manufacturing process to produce it.

In the years since 1980, Meade has continued to expand and improve its product line, adding numerous innovative SCTs almost constantly. The company has especially become known as the leader in high-tech computerized Schmidt–Cassegrains. This approach culminated in the introduction in 1992 of the LX-200 series of SCTs, the first practical and affordable Schmidt–Cassegrain to implement a computerized drive system that allows the user to "goto" desired celestial objects at the press of a button.

Which is Better: Meade or Celestron?

This is the question beginners ask experienced Schmidt–Cassegrain users more than any other. And, honestly, it is not an easy one to answer. Though there are differences in the telescopes sold by Meade and Celestron, they are at heart very similar. I have found the optical performance of today's Meade and Celestron SCTs to be virtually indistinguishable, and I really would be unable to choose one or the other company as the leader in optics. Mechanically too, the major brands have a lot in common. Neither Meade nor Celestron produces what are considered premium telescopes (like high-end APO refractors costing $10–15,000 or more), but both sell instruments that are well designed and reasonably reliable.

So how to choose? Not only are the scopes similar in design, the various tiers of instruments, from the simpler introductory models to the top-of-the-line flagship telescopes, are also *priced* similarly. But if you take a closer look at Meade's and Celestron's output, certain general philosophical differences become evident, as a perusal of this book's detailed buyer's guide will show. Meade tends to rule the high-tech roost. Its robotic wonder, the LX-200, has more features and options than the competing Celestron goto telescope (the Ultima 2000). Even Meade's less fancy models are generally more easily adaptable to add-on computers and similar devices than Celestron's telescopes. Celestron? Its innovations tend to be more of an optical and mechanical nature, and its telescopes are usually simpler electronically. Many of Celestron's CATs have been very popular with astrophotographers because of their simple, yet very sturdy, mountings. Another possible decision-maker is aperture. Celestron's size range of 5, 8, 11 and 14 inches has not been exactly duplicated in the Meade products. The Meades are available in apertures of 8, 10, 12 and 16 inches. The disparity in the range of sizes is not a great one, but if you're looking for a telescope of a particular physical size and weight, it may make a difference.

Anatomy of an SCT

Up to this point we've been talking about the optical and mechanical characteristics of the Schmidt–Cassegrain in general terms. Now, however, we're going to take a detailed tour of the average SCT and take a closer look at both the tube (the optical tube assembly or OTA) and the mounting that this optical tube rides on. We'll be looking at an 8 inch scope, but larger and smaller instruments will be very similar to our example. Unless otherwise noted, what's said can apply to either a Meade or Celestron model.

Optical Tube Assembly (OTA)

The first impression of a CAT's tube is that it's *short and fat*; it is very compact due to the folded optical system employed by these telescopes. Most SCTs have a focal length of f/10. Ten times the diameter of an 8 inch would normally mean a focal length of 80 inches, but because of the optical tricks we've discussed, this long focal length is packed into a tube only about 16 inches long. The tube is made of thin-walled aluminum and is usually painted with a nice glossy finish on the outside (black, blue, orange and white are some of the colors the CAT companies have used over the years). The inside of the tube is painted flat black. After the telescope's compact tube, the next thing to draw our eye is undoubtedly the SCT's big glass corrector plate (see Figure 3.2).

The corrector is recessed inward by about an inch from the end of the tube and looks more like a clear piece of glass than it does a real lens. In fact, many beginners mistakenly think that the only function of the corrector is to hold up the secondary mirror mount and to seal the OTA against the intrusion of dust. But as we know, this thin glass plate (about 5 millimeters thick) is a critical part of the telescope's optics. In the middle of the corrector plate and extending through it is the mounting for your telescope's secondary mirror.

Figure 3.2. SCT corrector plate.

The secondary mounting (or *holder*) both supports the SCT's small convex amplifying mirror and allows it to be adjusted to bring the telescope into proper collimation. Three screws equally spaced around the circumference of the holder allow the mirror to be tilted in one position or another until it is aimed in precisely the right direction with respect to the telescope's main mirror.

Moving around to the back of the telescope, you'll find the rear cell (see Figure 3.3). The rear cell holds the SCT's primary mirror, and provides a way to attach

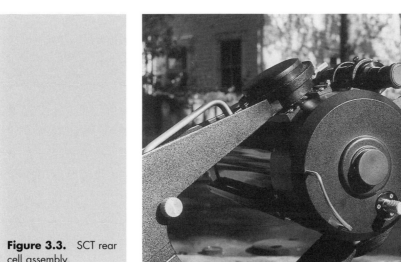

Figure 3.3. SCT rear cell assembly.

eyepieces and other equipment. The rear cell also includes the control for focusing the scope and is equipped with accessory holes for the attachment of a finder scope, a piggyback camera mount, counterweights and other external items.

In the center of the rear cell is a hole, the *rear port* of the SCT. It is surrounded by a raised and threaded metal lip. The size and threading of the rear port on both Meade and Celestron telescopes are the same, allowing items designed for one scope to be used on the other. Some equipment, such as camera adapters, threads directly onto the port, but most items require the installation of a *visual back*. The visual back is a ring-like adapter that makes it possible to insert American Standard 1.25 inch barrel diameter accessories like eyepieces or a diagonal mirror into the telescope. The visual back includes a set-screw for holding eyepieces securely.

Putting your eye to the visual back and looking through the rear port, you'll find yourself peering through the rear cell and mirror. You'll notice that rather than just looking out through the central hole in the main mirror, you're looking through a long tube. This tube, which is mounted to the rear cell and which extends through the mirror's central hole, is the *baffle*. In the Schmidt–Cassegrain this baffle tube performs two important functions. It helps prevent contrast-destroying *skyflood*, a tendency for light entering the corrector plate at oblique angles to bypass the telescope's mirror system and flood directly into the SCT's eyepiece. This would really hurt image quality, especially when observing bright objects like the Moon. The baffle keeps these stray light rays from intruding into our eyepiece. This tube also has another important job to do for the SCT. It provides something for the main mirror to slide up and down on.

If you're used to other types of telescopes, the focusing method used by the Schmidt–Cassegrain may seem shockingly different at first. Most instruments, refractors and reflectors, achieve focus by moving the eyepiece in and out. The focusing method on Meade and Celestron Schmidt–Cassegrains works differently. Look at the back, the rear cell, of your SCT, and you'll see a knob placed about halfway between the rear port and the edge of the telescope's back. This knob is attached to the actual main mirror holder inside the SCT by a threaded rod and mechanical linkage. Turning this control causes the primary mirror to slide up and down on the baffle. Turn one way and the mirror moves toward the back of the scope. Turn the other way and it moves up toward the corrector. Moving this mirror adjusts the position of the telescope's focal point. Turning this control allows the user to bring the image to exact sharp focus for a given eyepiece.

Why don't the SCT makers just do what other telescope builders do and arrange a focuser which moves the *eyepiece* in and out? There are a couple of advantages to the SCT system. Not having a moving eyepiece focuser makes for a more stable mounting for a camera. Heavy equipment can be attached to the SCT's rear cell, either via the visual back or by being threaded directly to the rear port. No worries about a heavy camera slipping and causing focus to shift during a critical exposure. A moving mirror system also means that the focal plane can be moved to a position far to the rear of the telescope by moving the mirror up the tube toward the corrector. This ensures just about any accessory you can put on an SCT can be brought to focus no matter how far back it sits from the rear port. This focusing system gives the SCT a generous amount of *focus travel*.

What else does the rear cell of an SCT do? It also supports the finder scope, the small telescope that helps you locate objects. A big telescope like an SCT, even when used at very low power, has an extremely narrow field of view (often considerably smaller than the width of the full Moon). This would make trying to locate objects very difficult without some kind of aid. If you don't believe this, go outside with a telescope some night and try to find even a bright star by sighting along the main scope's tube or slewing around! The finder provides a wide field of view (usually offering a magnification of about 6–12). This little telescope is equipped with a set of cross-hairs as in a rifle's telescopic sight and is mounted in a pair of adjustable rings so that it can be aligned with the main telescope. With a properly adjusted finder telescope, whatever appears in the finder should also be in the "big eye," making object location easy.

Move around to the front of the telescope again, look down through the corrector plate, and take a peek at the primary mirror. Like the secondary mirror, the primary should remain in pristine condition due to the sealed nature of the SCT tube and should rarely require any kind of user intervention. Theoretically, an aluminum-coated first surface mirror like a CAT primary might possibly need recoating in about 15 years (many Newtonians need to be resurfaced in less than 10 years because they're exposed to the elements). This is rarely necessary in the real world, though. The coatings are well done, and due to the protected position of the primary may last a lot longer than 10 or 15 years.

Mountings

What good is a telescope without a decent mounting? Absolutely no good at all. Without a sturdy, well-made mount to ride on, a telescope, even one with the very finest optics, is useless. The Schmidt–Cassegrain's mounting performs several very important functions. It supports the OTA and allows it to be aimed at objects of interest. Equipped with motors, it counters the turning of the Earth, tracking the stars and keeping celestial objects in the CAT's field of view. It can provide support for cameras, and furnishes the control systems necessary for the very fine corrections needed when taking long-exposure astrophotos.

All currently produced SCTs feature *equatorial mounts*. With an equatorial mount, a celestial object can be followed and kept in view with only one motion. Unlike the equatorial style mount, an *alt-azimuth mounting* (often seen on small, inexpensive refractors) requires *two motions* to follow a star in its progress across the sky – it must be moved in altitude and azimuth, up and down or side to side, as its name suggests. Over the centuries of the telescope's evolution, many different designs for the equatorial mount have been tried, but SCT manufacturers have settled on two designs: the fork and the German equatorial.

Fork-mounts

The fork-mount is very simple, but it is very adaptable and provides a neat, convenient and steady mounting for the average SCT. The fork, the style of mount chosen for the first C8s by Tom Johnson, is still the most widely used style of telescope support for Schmidt–Cassegrains. As the name describes, this mounting is formed from a large metal fork that is attached to the OTA on either side of the tube. The

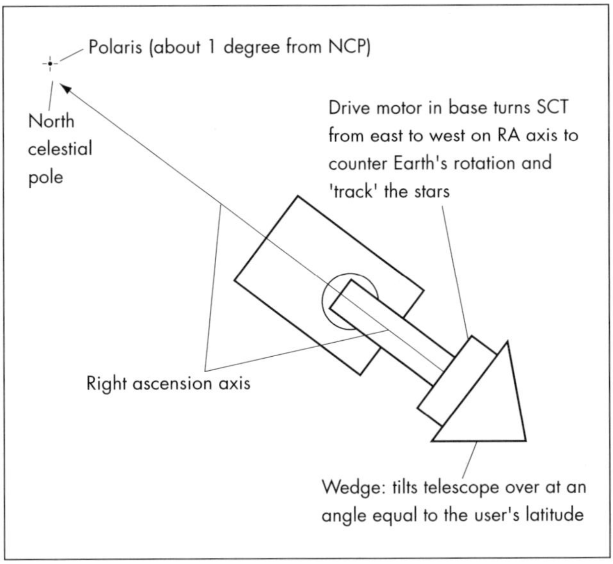

Polaris (about 1 degree from NCP)

North celestial pole

Drive motor in base turns SCT from east to west on RA axis to counter Earth's rotation and 'track' the stars

Right ascension axis

Wedge: tilts telescope over at an angle equal to the user's latitude

Figure 3.4. A wedge allows an SCT's right ascension axis to be pointed at the North Celestial Pole.

scope is free to move up and down in this fork thanks to two side bearings. This big fork sits atop a rounded base, the telescope's *drive base*, which can swivel 360 degrees. A motor mounted in the base can be engaged and will drive the telescope at the precise rate required for star tracking. This is the essence of the fork-mount. But just placing this combination of fork and telescope on a tripod wouldn't constitute an equatorial mount. We'd have a simple alt-azimuth mount that would require movements in two axes to track. Another component is required in order to make the fork into an equatorial. This is the wedge.

The idea of the wedge is a simple one. It allows the fork to be tilted over (see Figure 3.4) so that it can be pointed toward the North Celestial Pole. With the scope tilted over like this, the center of rotation of the telescope on its base forms the right ascension (RA) axis, the motion that corresponds to movement in celestial longitude. The movement of the scope in its side bearings now becomes movement in declination, celestial latitude. The combination of scope, mount and wedge is placed on a tripod, and the RA axis is aimed toward the North Celestial Pole. When the motor in the drive base is turned on, we find that it almost magically keeps objects in view for as long as we care to look at them (assuming an accurate polar alignment, the pointing of the RA axis at the North Celestial Pole, has been accomplished). Let's take a detailed look at fork-mounting.

The fork's side bearings attach the telescope to the mount and allow for movement in declination. These are fairly large, several inches in diameter. At least one of the bearings is commonly equipped with a scale graduated in degrees and a pointer. As the telescope is moved, the pointer, which is attached to the fork arm, remains stationary, and the scale moves indicating different values of declination. This is a setting circle. With a properly aligned telescope it can be used to help point at a particular object by moving the tube until the proper value of declination is indicated (the object's declination value would have been looked up in a book or list). One of the side bearings also includes a mechanical lock for securely locking the telescope's declination axis. With this lock engaged, the telescope's declination slow-motion control can be used.

The slow-motion control is located at the base of one of the fork arms (see Figure 3.5), usually the one equipped with the declination locking lever. It is a simple threaded bolt that runs through the fork arm, and sports knobs at either end. A threaded block rides on this screw and is attached to a tangent arm that runs vertically along the inside of the fork arm. Turning one of the slow-motion knobs moves the tangent arm forward and back. The upper end of this arm is connected to the side bearing. When the arm moves, it causes the scope to move in

Figure 3.5. The declination slow motion control is the silver-colored knob at the far left, at the base of one of the fork arms

declination slowly and smoothly, aiding greatly in positioning objects in the field of view, especially at high power. The declination slow motion only works when the declination lock is at least partially engaged.

The telescope's drive base forms the right ascension axis and provides a home for the drive motor and any onboard electronics or computers. Like the declination axis, the right ascension axis has a graduated setting circle, this one marked in the hours and minutes of celestial longitude. This setting circle is attached to a round plate that is part of the fork. As the fork is rotated and the RA circle turns with it, a pointer on the stationary drive base indicates the telescope's current right ascension.

Also located on the circular plate the fork rides on is the scope's right ascension lock. This secures the telescope on the RA axis. With the RA lock engaged, the SCT's drive motor is also engaged and turns the telescope through a gear system. As the scope tracks, the setting circle remains stationary and continues to indicate the proper RA of an object as it moves across the sky. With the lock disengaged, the setting circle turns with the telescope to indicate the changing right ascension of the SCT.

Adjacent to the RA lock is the right ascension slow-motion control. Like the declination slow control, this allows the telescope to be moved at slow speeds by hand. This control works slightly differently from the declination slow motion, however. For the declination slow motion to work, the lock must be on – engaged. To use the right ascension slow motion, though, the RA lock must be at least partially off – *released*. This is *very* important, as attempting to turn the RA slow motion with the lock engaged *can cause severe damage to the control's gearing*.

The SCT's control panel may be found on the side of the drive base – a little inconvenient as it makes the switches and indicators a bit inaccessible – or on the top of an extension to the drive base as in Figure 3.5. This panel may be very simple, featuring just an on/off switch and maybe a socket for a remote hand controller; or it may be festooned with numerous dials, switches, lights and sockets required by a modern computerized SCT. In addition to a power switch and a hand controller input, some features commonly seen on modern Schmidt–Cassegrains are inputs for electric declination motors, switches for controlling the speed of the telescope's drive, lights or meters indicating the status of a battery power supply, and outputs for 12 volt accessories.

Let's look *under* this panel, too. You open the drive base on a Celestron telescope by removing the control panel. On most Meade telescopes you'll remove the bottom of the drive base to look inside. One important item found inside the base is a battery holder for telescopes that use built-in batteries. You'll also find the brains of your telescope – the SCT's electronics assemblies. These printed circuit boards (PCBs) may be simple and may be used only as motor controllers – *drive correctors* – or they may be highly complex, forming a computer motherboard.

Visible inside the drive base of all SCTs will be the electric motor and gears that comprise your CAT's clock drive (see Figure 3.6). In its most basic form, a telescope drive does one thing: it turns the scope on its mount at the sidereal rate. This is one revolution every 23 hours and 56 minutes. This *sidereal rate* is equal to one revolution of the Earth and is what's needed for tracking the stars and other objects. By turning from east to west at this rate, the telescope drive counters the turning of the Earth and keeps celestial objects in view. The electric motor used by your telescope will be one of three different types. The simplest is a simple AC synchronous motor. This is accompanied by a gear system that provides the sidereal rate.

Figure 3.6. The base of the SCT contains the scope's drive gearing and electronics.

The operation of an AC drive is simple. You plug it into a household electrical socket and it runs. Unplug the scope and it stops. The speed of an AC motor of this type is usually quite accurate because it is *synchronized* to the frequency of your AC current. For operation in the field, it can be powered from a 12 volt storage battery by means of an inverter, which converts DC from a battery into AC to run your telescope. If you plan to do astrophotography, a very special kind of inverter, a *drive corrector,* is required. This allows the photographer to make the very small corrections to the aim of the telescope needed during long-exposure photography.

The drive corrector is a type of inverter and is connected to a DC source, a 12 volt battery usually. DC from the battery is converted to AC that the telescope motor can use. But the corrector includes an additional circuit, one which varies the frequency of the alternating current output when a button is pressed. Varying the frequency of AC current up or down speeds up or slows down your synchro motor for photographic guiding. Most drive correctors feature wired remote controls. These *hand paddles* are equipped with at least two push-buttons that correspond to east and west. Some drive correctors also have DC outputs for small declination motors and buttons for north and south as well. With the addition of an add-on electric declination motor, these drive correctors can move the telescope north, south, east or west when the appropriate button is pushed. Don't think that a drive corrector will allow you to slew your scope across the sky, though! The highest speed a drive corrector can deliver is only a couple of times that of the sidereal rate. It would take *hours* to move the scope far. These devices are *only* suitable for small photographic or centering corrections.

As the 1980s wore on, Meade and Celestron decided their SCT buyers would like a telescope that eliminated the need to carry around inverters or drive correctors, and before long DC-powered telescopes appeared. Some of these still required the use of an external battery. But many models were being powered by nothing more than a tiny 9 volt transistor radio battery. Most of these telescopes used DC servo motors, which proved to be efficient and reliable. Celestron, for example, introduced several new SCTs that would run for as much as *50 hours* off a single small 9 volt cell. Even telescopes still requiring a separate battery were able to offer a host of new features with the coming of DC scopes. One of the most important of these features was the inclusion of various drive rates and slewing speeds.

The built-in drive correctors (actually electronic control circuits rather than real drive correctors, but that's what amateurs initially called them) could be set for the normal sidereal rate, solar rate, lunar rate and other specialized drive speeds. The accuracy of the telescopes' tracking was also quite good, with quartz crystal oscillators providing a reference. You'll still see reference to quartz telescopes. But this has nothing to do with quartz glass optics. The introduction of DC telescopes was a necessary prerequisite for the development of the computerized goto telescopes of today.

Today, even non-goto telescopes incorporate a lot of computerization into their drive mechanisms. In most of them the original servo motors have been replaced by *stepper motors* (the same type of motors used in computer printers). Unlike a servo motor, or most other electric motors, these devices rotate in small, discrete increments – steps. These motors are controlled by small, onboard dedicated computers.

Despite the advanced electronics used in modern telescopes, they still require old-fashioned mechanical gears to transfer motion from the motor to the telescope itself. The earliest systems, including those found in the first C8s, used *spur gear* drives. In the spur gear systems, a small gear with straight teeth attached to the motor drove a larger gear of the same type coupled to the telescope. The ratio of the two gears was arranged so the result was a sidereal rate of turning. Spur gear systems work fairly well, are inexpensive and can be highly accurate in speed. Their drawback is that the nature of the gears' teeth causes tiny random variations in speed. This is not a problem for the visual observer, but it means that the astrophotographer must monitor the scope's aim all the time and be ready to push a drive corrector button when these random variations show up.

In the early 1980s, Meade introduced worm gear drives with its first SCTs. This type of gear set utilizes a worm gear, a cylindrical, helically toothed gear, to drive a large gear, a *worm wheel*. The helical nature of the worm and the more precisely cut teeth of the wheel result in a very accurate drive rate. Much of the spur gear system's randomness is eliminated. Like any mechanical gear system, worm gear drives still show some error. But this is usually *periodic error*, a slow and regular variation easy for the astrophotographer to guide out using the scope's hand controller.

The German Equatorial Mount

So, the fork equatorial is the best way of mounting a Schmidt–Cassegrain telescope? The manufacturers seem to think so. Most of the SCTs sold over the last 30 years have come equipped with this kind of mount. Undeniably, the fork can work well for the CAT user – if it couldn't, it would have disappeared a long time ago. It is an extremely compact system, very convenient, versatile and user-friendly, but it does have a few limitations.

The first problem the new fork-mount owner encounters occurs when the telescope is pointed at a celestial object in the far northern part of the sky. While pointing north, a polar-aligned fork type SCT has the rear end of the telescope positioned between the fork tines and close to the drive base, making it very difficult to find a comfortable position for viewing. Bearable, but uncomfortable. The closer to Polaris and the celestial pole the scope is pointed, the worse and more awkward things become. This is especially problematical for the deep sky photographer. Cameras and adapters can extend far enough out from the back of the telescope to make it impossible to point the scope very far north. When the photographer moves the back end of the telescope between the fork arms, the camera bumps into the drive base.

A further criticism of the fork is its size and weight. An 8 inch telescope in its mounting is reasonably light and manageable. A 10 inch becomes a problem weight-wise, and a 12–14 inch *demands* a permanent installation. The problem is that the fork-mounted telescope's OTA cannot be easily removed from its mount. The smallest component of the disassembled telescope includes the OTA, the fork and the drive base. With care, thought, planning and help, large SCTs can be used in the field (I've seen people drag 16 inch CATs to star parties!). But this is certainly not something to undertake lightly. A large fork-mount telescope may also discourage frequent use – who wants to set that monster up in the backyard for a quick half hour weeknight session?

The fork has also been criticized on the grounds that it is prone to vibration. How this can happen is easy to see when you step back and think about what you have with a fork-mounted telescope. A large mass is suspended between two relatively thin fork tines. It doesn't take much for these arms to start acting like the tines of a tuning fork – vibrating and spoiling your high power view or your photograph. In practice, most modern fork-mounts are pretty well designed and minimize the vibration problem. But the potential for shakiness is still there. This is made worse by the basic off-center/off-balance arrangement of the telescope. Tilting an SCT with a wedge for equatorial use just increases this propensity for vibration.

But there is an alternative, the *German equatorial* (see Figure 3.7). This mounting, which, like the fork, is designed to track the stars, has a lot of fans among photographers and other serious amateurs. The German equatorial mounting, or GEM, has been offered as an option by manufacturers of SCTs almost since the first CATs rolled off the assembly line. While it has not and probably will not displace the fork-mount as the most popular support for the SCT, it does have some advantages that make it worthy of your consideration.

At first glance the GEM looks complicated, but the design is really simple and elegant. Referring to the picture, you can see the GEM is composed of two axes mounted perpendicularly to each other. One

Figure 3.7. The German equatorial mount is a common alternative to the fork mounting for SCTs (photo courtesy of Meade Instruments Corporation).

axis is attached to the tripod and forms the mounting's right ascension housing. A shaft runs through this housing and is the actual RA axis of the telescope. It is attached to the *declination housing* which is at a right angle to the RA axis. This declination housing also has a shaft running through its center forming the scope's declination axis. One end of this shaft is connected to the telescope. The other end extends through the declination housing and has one or more counterweights attached to its end to balance the heavy telescope tube mounted on the opposite end of the shaft. That's a GEM in a nutshell. But let's dig into this contraption a bit deeper, starting with the RA axis.

As we've learned, the RA axis of a telescope is pointed at the North Celestial Pole in order to align an equatorial for star tracking. To this end, the GEM's RA housing is mounted on a strut. This strut contains a bolt or lock knob that can be loosened, allowing the whole RA housing to be tilted upward to point at the altitude of the North Star. Most GEMS provide a means for this strut and the RA housing to be rotated from side to side, enabling the RA axis to be aimed at the azimuth of North or South Celestial Pole during polar alignment.

The RA drive gear is attached to the right ascension axis. If a motorized clock drive is added to the mount, it will turn this gear, moving the scope slowly in right ascension, allowing the telescope to follow the stars. The better and larger this gear is, the better your tracking will be. The RA drive gear is the place where the smaller German mounts, offered as an option by CAT manufacturers for their 8 inch scopes, lose out to fork-mounts. The gears on the fork-mounts are larger (the big drive base makes a convenient housing for a big gear), often making for smoother, less error-prone tracking.

Like their fork-design cousins, GEMs are equipped with slow-motion controls. These consist of knobs attached to the RA and declination worms. Turning these knobs moves the scope slowly in the chosen axis. For convenience, the slow-motion knobs are often placed on the ends of long shafts or, in the case of inexpensive mounts, flexible cables. These flexible cable style controls should be avoided if possible, since the springy movement caused by touching the cables imparts vibration to the scope.

Moving to the declination housing, you'll find a declination setting circle and pointer that work similarly to the declination circles found on the fork-mounted telescopes. The declination shaft is contained within this housing. One end of the dec shaft is equipped with a bracket and dovetail arrangement that allows the scope OTA to be easily attached and removed from the mounting. Unlike the fork-mount, it is normal for the telescope optical tube to be removed from the GEM mounting for transport and storage. The opposite end of the dec shaft extends far enough out of the dec housing to allow the counterweights mounted there to slide up and down for fine balancing of the scope on the RA axis (you balance the telescope in declination by moving the OTA forward and back in its dovetail bracket). The end of the declination shaft is usually equipped with a safety bolt that prevents the heavy counterweights from accidentally sliding off the end of the dec shaft. This is a very good thing. Your author, as a very young man, had a bad experience with a counterweight on a GEM-mounted 4 inch reflector that didn't have this feature. In the course of adjusting the scope's balance, I let go of the heavy counterweight, which I'd loosened to balance the scope, and it slid right off the shaft and onto my big toe!

In order to track the stars, an equatorial mount needs a motor. The area of motors and electronics is one where the GEM really shines. In this regard, the German equatorials are much like "clone" computers – *you* decide what kind of features you want on them. Most GEMS offered for sale with SCTs don't come with any motors or electronics *at all*. Initially, this was to help the SCT companies offer consumers a working telescope for less than the cost of their most basic fork-mounted models. Being able to list the price of the scope without a motor made for an attractively cheap quote. But in some respects this has worked to the *advantage* of the amateur astronomer. You can choose how much money to spend on your telescope's electronics. Are you a stay at home type? A backyard observer doing visual observing? Then you can put a simple and inexpensive AC drive on your scope. Want a telescope that automatically points at objects, just like a goto fork-mount? Mount makers offer robotic drive kits for your GEM too.

The GEM mount is a viable alternative for the SCT fan, but one that is just now really coming into its own. Both Celestron and Meade now offer big German mounts that are exceptionally stable and full featured.

Celestron's model is intended more as an astrophotography platform (serious astrophotographers usually prefer GEMs) while Meade's current offering includes a computerized goto option (Meade seems to be mainly pushing this GEM for its refractors right now, though).

Tripods

Just as no telescope can be considered complete without a decent mounting, no mounting is complete without a good, solid tripod to support it. The tripod is critically important in the quest for a steady CAT, especially if photography is the goal. In the beginning, Celestron's original C8 was available with the tripod that arguably provided the best combination of lightness and stability ever seen in a Schmidt–Cassegrain. This was the famous *triangle tripod*. The legs and braces were a series of triangles forming, from an engineering standpoint, the perfect design for a tripod. A similar arrangement is often found in the high-priced tripods used by professional photographers for large format cameras or by surveyors for their expensive instruments. This tripod provided a very nice platform for the C8, but users didn't like it and complained that the tripod was not adjustable or collapsible. Most owners described loading the triangle tripod into a small sedan as being akin to wrestling with an octopus!

As a result of consumer reaction to the original tripod, all Celestron telescopes have featured collapsible units with extendable legs ever since. Meade's products have followed this same design. The usual tripod found on a CAT is a tubular affair (see Figure 3.8), with legs that can be extended to bring the height of the tripod head to about 4.5 feet; most also feature a *spreader*, a metal bracket that fits beneath the tripod head and has three extensions which push against the legs. A threaded knob and bolt allow the spreader to be tightened against the tripod legs ensuring they are held firmly apart. The Celestron tripods have varied a bit in quality over the years, but have been fairly solid, if not nearly as light and stable as the original triangle tripod. The same thing has been true of the Meades: some variation over the years, but mostly good-quality extendable-leg tripods.

Figure 3.8. Most modern SCT tripods feature tubular extendable legs.

Our tour of a generic SCT is at an end. We've already mentioned a few specifics applying to various manufacturer's telescopes. But in the next chapter we'll really get down to specifics and review the large variety of new SCT models. There is a confusingly large array of CATs for sale these days, but this confusion can be lessened if you think back to our basic CAT optical tube on a simple fork or German equatorial mount. Under the skin the many SCTs produced over the years are almost identical, with the differences being largely cosmetic or in the form of numerous options.

Chapter 4

Which CAT?

Choosing a New SCT

Amateur astronomy is growing as never before. Combine with this the fact that the SCT is now undeniably the most popular style of telescope for serious observers, and it's no surprise that Meade and Celestron provide a wide variety of SCT models for amateur astronomers. An increasing number of similar telescopes, mainly the SCT-like Maksutov–Cassegrain telescopes (MCTs), is also being sold by other manufacturers. This large number of seemingly similar CATs can be confusing and daunting for the first time buyer. But it also means there's probably a Schmidt–Cassegrain out there that will be perfect for you, *your* observing interests, and *your* budget. A little study and comparison can make buying your first telescope a relatively painless affair.

If you're considering buying an SCT or other catadioptric telescope, you need to decide how much you can afford to spend on a new telescope and exactly what you plan to do with it. Price-wise, the SCTs have fallen into three groups: introductory bargain scopes, medium-priced models and top-of-the-line "luxury" CATs. The differences among these groups are in their mounts and electronics; both Meade and Celestron use the same tube assemblies for all their models. Just as the scopes fall into three tiers of price, they also seem to be aimed at three classes of astronomer: general interest/beginner, serious amateur/astrophotographer, and computer fan/specialized interest worker. Which group do you fall into?

Entry-Level SCTs

These are the lowest-priced telescopes in both manufacturers' product lines. They may be either fork-mounted or provided with light or medium-sized German equatorial mounts. Formerly, all the fork-mounted scopes in this group came with simple AC synchronous motors. But this is a thing of the past. Due to "feature creep," the desire of the manufacturers to offer as many features as they can to remain competitive, the clock drives on these telescopes are all DC powered now. In my experience, they are *easily* as good as those on the medium or top grade SCTs of only 10 years ago. The main thing that distinguishes these CATs from the more expensive models today is the lightness of their mountings and tripods. The accessories provided with these telescopes, finders, eyepieces and star diagonals, also tend to be of a basic nature.

Is a bargain CAT the right telescope for you? It may be. These SCTs are particularly suited for astronomers who are interested in visual observing. Especially city dwellers who must do most of their observing from remote dark sites. These scopes are light and are very easily transportable even by lightly built observers. Like all SCTs, they are quite versatile, and are just as good for touring the lunar landscape as they are for exploring deep space. The drives and electronics featured by *all* modern introductory SCTs are more than adequate for visual use. In some cases they may offer more features than will ever be used by the average visual observer. These telescopes are certainly *not* optimum for photography or CCD imaging, mainly because of their light mountings, but they can be used to get your feet wet in astronomical picture taking and can produce some genuinely wonderful astrophotos.

Who should stay away from these least sophisticated SCTs? Mainly those folks who *know* they are very interested in photography and want a telescope that is at least partially intended for this purpose. If you have your heart set on beautiful long-exposure photographs of galaxies and nebulae, you'd do well to *start* out with a medium-grade SCT. A telescope with a better mount and drives will lead to much less frustration for the new imager. But what if you just *have* to have an SCT now, but know you'll want to take photographs and really can't afford one of the more expensive models? If you can't bring yourself to save for just a while longer, one of the introductory GEM mounted scopes may be a good choice. These scopes can often be had with *usable* German mounts for only a little bit more money than OTA-only SCTs (no mount) command. Once the user is ready to move up to a more sophisticated setup, a top-of-the-line GEM can be purchased and the OTA easily switched over.

Mid-Level SCTs

The most distinguishing characteristics of these telescopes are their upgraded mounts, wedges and tripods, all of which are substantially heftier than those found on the entry-level scopes. But many other details have also been improved. Finder scopes are often bigger, and the supplied eyepieces and star diagonals are usually of higher quality. Most importantly, the clock drives on this tier of instruments are better in every way. Sometimes they are more accurate and almost without exception they are more versatile. In addition to selectable drive rates (speeds) this group will often include special drive features like PEC (periodic error correction), which is designed to furnish the drive accuracy needed by photographers.

Who should pay the extra money needed to buy a mid-grade CAT? Photographers should, certainly. The more sophisticated mounts and drives of these CATs are really vital for success in imaging. Photography through the telescope is hard enough without making it more difficult with the light mounts and more error-prone drives of the less expensive telescopes. This is not to say that this level of telescope is *only* suitable for photographers. If you are a committed planetary observer, for example, you may find that the steadier mounts found on these telescopes make high-power viewing of the Moon and planets more practical and pleasant. Due to their significantly upgraded mountings, these telescopes are noticeably heavier than the basic models, but most observers will still find the 8 inch models portable enough for easy transport to dark observing sites. The mid-level

Schmidt–Cassegrain is an attractive telescope for the amateur who wants quality and real photographic capability, but who is really not interested in getting involved in computers.

Luxury SCTs

The top tier of SCTs can actually be divided into two groups: the computer scopes and the big GEM photo scopes. The most popular of these luxury telescopes, though, are undoubtedly the computerized goto models that find celestial objects at the push of a button. The big German mounted CATs are aimed at a substantially smaller audience of experienced and advanced amateurs.

Fork-Mount Goto SCTs

Everybody, it seems, has an opinion about goto scopes, those modern technological marvels that take all the work out of observing. Want to view an obscure galaxy? Turn on your telescope, go through a short and simple alignment procedure, and punch the catalog number of your desired galaxy into the computer hand control of your LX-200 or Ultima 2000. The telescope whirs, moves across the sky and places the object in the field of view. The manufacturers have refined the electronics and mechanics of these telescopes until they are capable of truly astounding accuracy. So precise are these telescopes that they can place an object close enough to the center of the telescope's field so that it is within the area of the tiny imaging chip of a CCD camera. These telescopes are also easily adapted for control by external computers. It is now common to see one of these scopes working in conjunction with a laptop PC. The computer displays a chart of a star field, the user moves the mouse pointer to an object of interest, and the telescope points at that object.

Who's a good candidate for goto ownership? People who are fascinated by and comfortable with computers and technological gadgets make up a large segment of the target audience. Observers who are conducting special programs like supernovae searches or who are involved in scientific data gathering like variable star monitoring are also likely to benefit from these CATs' amazing capabilities. If you need to check 30 galaxies for supernovae every evening, star hopping to your destinations using your finder scope and star charts isn't fun – it's a time-waster.

Who should forego these highest-tech telescopes? If you don't like computers, get glassy eyed at the mention of the word "Internet" and enjoy the simple side of stargazing – the sounds of nature rather than the beeps and buzzes of a goto scope, you're not likely to have much fun with an LX-200 or an Ultima 2000. It's not easy, I know, to pass up one of these marvels. The SCT makers naturally devote a large share of their advertising budgets to promoting these telescopes. They are made to look very attractive, and the idea that computerized pointing is indispensable is at least implied. But for some of us, slower and simpler is better. Observing is not a race. In my opinion, it's much more rewarding to devote an evening to observing a few objects in detail – really appreciating their majesty – than to merely glance at 20 or 30 galaxies, not seeing much more than one smudge after another.

There's also the question of reliability. Meade and Celestron have gone to great lengths to ensure that their computerized telescopes are trouble-free. They have also done a good job of seeing that the telescopes get fixed quickly and properly when something goes wrong. That's a good thing, because it's almost a *certainty* that something will eventually go wrong with a goto-style telescope. These are *very* sophisticated instruments which incorporate numerous highly complex mechanical and electronic assemblies. They have to do a lot, and their prices have to be kept within the reach of the average advanced amateur astronomer. Because of these facts, there is simply no way they can be expected to be as reliable as, say, a basic-level SCT whose drive consists of a simple battery powered electric motor. If, for any reason, you cannot stand the idea of having your telescope out of service for maintenance, you should probably pass up these most sophisticated and complicated CATs.

Many long-time amateur astronomers would also tell you there's another reason for avoiding a goto scope if you're a beginner. According to them, having a scope that automatically points to objects of interest discourages the new astronomer from learning about the sky. This person will (according to some old hands) quickly get bored with facile observing. There's no challenge to finding objects with this type of telescope. Also, it's said that the goto observer doesn't establish any kind of a relationship with the night sky. It remains a mystery, and many of the beautiful marvels that the average observer stumbles on by accident are lost to the computer person. I can agree with this sentiment to some extent. I think that learning the constellations is one of the most wonderful things a person can do. In my own case, I also find the *hunt*, the search for galaxies, star clusters and nebulae, to be almost as rewarding as looking at them. But, on the other hand, amateur astronomy is supposed to be fun. I've seen many new observers drop out because they couldn't get over that initial hurdle of learning the sky and learning how to locate objects. With a goto scope, the new amateur can start looking right away. Nothing says he or she will never learn the sky, either.

One of the more recent developments in amateur astronomy is the spread of goto mountings from high-priced telescopes to entry-level models. Some of these telescopes may work quite well, but maximum accuracy and reliability are still confined to the top-tier CATs.

Luxury GEMs

The other group of top-of-the-line scopes is the Big GEM-mounted SCTs. These telescopes are mounted on very high-quality German equatorial mounts. The GEM systems of this type from Celestron have proven to be quite popular with advanced users, because the buyer can obtain a turn-key system – a complete scope setup ready to go for advanced picture taking. Celestron currently has available an 11 inch and a 14 inch SCT in this configuration. But a system of this type can also be put together by buying an SCT OTA from Meade or Celestron and a GEM from any one of a number of manufacturers. Although the class of GEM mountings usually found on these scopes feature high-quality drives with all the latest features, until recently they have not featured the goto options found on the top-of-the-line fork-mount scopes. This is changing, however, with at least one precision GEM maker, AstroPhysics, marketing a highly sophisticated computerized goto GEM mount.

How Big?

This brings up the question of size. Beginners who've done their pre-telescope buying homework soon get the idea that the bigger an aperture a telescope has, the better (see Figure 4.1). This is *true* as far as scope performance goes. A large-aperture telescope will almost always outdo a smaller one – on everything, Moon, planets, and deep sky. The larger a CAT's mirror, the better its resolution, that is, the finer the details that can be detected in an object. Even more importantly, especially when viewing the deep sky, larger mirrors deliver *more light*, so dimmer objects like faint galaxies and nebulae can be seen in a larger telescope. Light gathering ability goes up steeply as the mirror's diameter increases. A 10 inch SCT gathers around 50% more light than an 8 inch model. A 16 inch CAT delivers about four times the light of an 8 inch, allowing you to see stars down to around magnitude 15 (an 8 inch scope allows you to easily make out stars of about magnitude 13).

If an 8 inch SCT is good, a 12 inch is better and a 16 inch is the best, right? If all we're considering is the raw performance of a telescope, yes, that is true. But *looking through it* is not the only thing you do with a

Figure 4.1. A bigger SCT is not always better.

telescope. You also have to *carry* it into the backyard, load it into your automobile for trips to dark sites, and *pay* for it. All too often, I've seen new amateurs thumb their noses at the ubiquitous 8 inch SCT, deciding on a 10 inch or 11 inch instead. Why "all too often?" Because I often find, a few months down the road, that the wonderful new scope has taken up residence in a closet or has been sold. It was great at first. The views were spectacular. But after the first newness wore off, the owner started feeling a little reluctant to set that big scope up. It's just too big and heavy. There's an old saying, and a very true one among amateur astronomers: "The best telescope is the one that gets used the most." For the average amateur, this is more likely to be a "little" 8 inch than a huge 12 inch.

Small Scopes

You can go larger. But you can also go smaller. Both manufacturers produce Schmidt–Cassegrains (or Maksutov–Cassegrains) in sizes smaller than 8 inches too. 5 inch and 3.5 inch instruments are currently being marketed. With all the worry about size and weight, shouldn't these be even better than the 8 inch for the new stargazer? There's no denying that the portability of the smaller telescopes is a plus. A telescope which can be picked up tripod and all and easily moved outside on a moment's notice will get used *a lot*.

But I do have reservations about telescopes of this size. When the aperture of a telescope sinks below 6 inches, its performance quickly nose-dives. An example is provided by the great globular star cluster in Hercules, M13. In an 8 inch telescope used under reasonably dark suburban skies, this object reveals itself to be a huge ball of stars. You can stare at it in wonder for hours, literally, lost in its seeming uncountable Suns. In a 5 inch, though, M13 is just a blob. A bright blob, yes, but basically just a blob, with few stars resolved under suburban conditions. In a 3.5 inch telescope, the blob of M13 isn't even that bright anymore, looking more like a nondescript smudge. A small scope *can* do well on the Moon and planets (though not as well as an 8 inch), of course.

Should anyone ever consider a little scope? Of course, there are always exceptions. An urban dweller who has to carry her telescope down several flights of stairs to observe probably would be better off with a small scope. Physically challenged persons or very young observers for whom even an 8 inch is just too much might also be well served by the "little guys."

Which Focal Ratio?

As was mentioned in the first chapter, *focal length* is the distance between the telescope's objective (be it lens or mirror) and the focus point, the place where the image is formed. *Focal ratio* is the ratio of the focal length of your scope to the diameter of its lens or mirror. An 8 inch telescope with a focal length of 80 inches has a focal ratio of f/10 (80/8 = 10). f/10 is the most common focal ratio for SCTs, but some SCTs can be bought (from Meade) with an f ratio of f/6.3.

Why would you want to do this? A telescope with a lower focal ratio offers lower magnifications with all eyepieces than a longer focal length instrument. This means that for a given eyepiece an f/6.3 8 inch telescope will produce a lower magnification and a wider field of view than an f/10 CAT. This can be nice. The f/6.3 scope gives the nice wide views appropriate for big deep sky objects like large star clusters while still using normal focal length eyepieces.

Deep sky photographers also like short focal ratio telescopes. Not only do they produce wider fields. They require shorter exposures than longer focal length telescopes do. This is really nice, because a photographer must expend a lot of effort keeping the telescope tracking a guide star throughout an exposure. It's easy to see why shorter exposures are possible with lower f-ratio scopes. When taking pictures of an object, an eyepiece is not used to magnify the image produced by a telescope – the telescope *becomes* the camera's lens. The image it delivers to film is the lowest possible magnification possible for its objective size/focal length.

A lower focal ratio, and therefore lower magnification, telescope produces a smaller, brighter image of extended objects on the film frame than a longer focal length telescope of the same aperture. This rule only applies to *extended* objects: nebulae and galaxies. The pinpoint light sources of stars need the same exposure regardless of f-ratio. Because of this smaller, brighter image, a nebula or galaxy shows up on film a lot faster with the short focal length/low f-ratio telescope than it does with the long focal length/large focal ratio telescope. Like terrestrial photographers, astrophotographers refer to smaller f ratio telescopes as being *faster* than large f-ratio ones. An f/6.3 telescope is faster than an f/10 one. The image forms on film faster than it does with a slower f/10 SCT.

SCT Buyer's Guide

It's time to really pick a CAT. The buyer's guide to new SCTs that follows is current as of the writing of this book, but CAT manufacturers do, like computer, television, and car makers tend to change models and pile on features every year. Many models do hang on for many, many years, however. If you've given due consideration to the issues we've talked about in this chapter you should now be well equipped to use this listing to help you choose *your* CAT.

Entry-Level SCTs

Meade LX-10 8 Inch SCT

The LX-10 (see Figure 4.2) is currently Meade's entry-level fork-mount Schmidt–Cassegrain. As supplied, the telescope comes with a wedge, but the tripod is an extra-cost option. The telescope doesn't come with a case either, but Meade's soft case cordura unit can be ordered for use with this unit. Like all current Meade SCTs, the enhanced multicoated optics group, MCOG, optical coatings for the mirrors and corrector plate are now standard. The telescope is currently only available with f/10 focal ratio optics. If you want a Meade fork-mounted SCT, this is as low as you go. It's an economy model for sure, but one which is surprisingly well appointed.

I was a bit surprised when I finally had a chance to examine an LX-10 in detail. I expected to find a lot of plastic used in its construction given the very low price of these telescopes (the LX-10 costs about what I paid for my first SCT – in 1976). But this was not the case. The entire scope is well built and is very comparable in terms of quality to the other instruments in Meade's line. After

Figure 4.2. The Meade LX-10 8 inch (photo courtesy of Meade Instruments Corporation).

close examination, I also have no doubt that the OTA is, as Meade claims, the same one with the same quality of optics used on their higher-priced telescopes. The fit and finish of the LX-10 are very nice, though the instrument didn't seem quite as well done as some CATs I've used over the years (most of which cost about twice as much as the LX-10). The scope looks very good, being attractively finished in Meade's customary dark blue paint. Standard accessories include a 1.25 inch prism type star diagonal, a 26 millimeter Plössl eyepiece, a 30 millimeter finder, and a hand controller (the hand unit is, for some reason, optional on LX-10s sold in Europe) for the telescope's battery-powered drive.

My first impression of this SCT, beyond the fact that it was very attractive, was that the fork and drive base are a trifle smaller than I'm used to. After operating and using the scope briefly, though, I quickly came to the conclusion that they are not outrageously undersized. The LX-10 has a nice solid feel even though it is undeniably a lightweight and inexpensive CAT. While I usually preach the value of heavy-duty telescope mounts, this low-weight design does have its advantages. Even younger and smaller observers won't have much trouble carrying and setting up the LX-10. The tripod normally sold for the LX-10 is a lightweight model that lacks extendable legs and a stabilizing spreader but it does work fairly well for visual observing. There is no denying, though, that the standard tripod and the included lightweight wedge are undersized for photography. An LX-10 owner wanting to get involved in imaging would be wise to purchase one of Meade's heavier duty tripod/wedge combos.

I was able to get used to the lightweight wedge and tripod, but I never have gotten comfortable with the LX-10's finder telescope. The unit included with this SCT has a 30 millimeter aperture objective lens. While the finder is of decent quality, it is really just too small, especially for use in light-polluted areas where it's hard to find guide stars to help you locate deep sky objects that are invisible in any finder.

Surprisingly, this inexpensive CAT is supplied with a very good eyepiece. In recent times, scope makers have tended to scrimp here, shipping their lower-priced SCTs (and even some of the costlier ones) with basic and fairly low-quality oculars. The simple Kellner eyepiece has been by far the most popular eyepiece shipped with recent telescopes. The LX-10, in contrast, comes with a very high-quality Meade Series 4000 Plössl.

How well does this little CAT *work?* I've been able to give this scope a very thorough evaluation, and my report is a favorable one. When properly collimated, these telescopes produce images, on both planets and deep sky objects, which are *very* pleasing, comparing quite favorably with those I'm accustomed to in much more expensive telescopes. The planets are always a severe test of a scope's optical quality, but the LX-10 acquitted itself quite nicely on the Sun's family. In moments of good seeing, Saturn is a marvel in this humble scope. Even though the rings were still not fully "open" when I did my initial evaluation of the LX-10

(1999), a gratifying amount of detail was visible in them. Cassini's Division was obvious and easy to see. The subdued crepe ring was also visible without too much straining. On one *really* good night, I was able to increase magnification to over 400×, and was rewarded with an image which not only did not fall apart – fuzz up – but which continued to deliver detail: brightness variations across the rings and considerable cloud banding on the disk of Saturn.

The battery-powered stepper-motor drive (which is usually run with a set of four internal AA cells) on this scope works quite well. While it, like the drive of *any* scope, shows some periodic error, it is not outrageously large, and appeared to be smooth and regular. I felt that this scope would be quite easy to guide during long exposures. What really shocked me, given the economical nature of the LX-10, is that the drive features a worm gear drive system. A spur gear set is more commonly found on less expensive telescopes.

The LX-10s sold in the US are equipped with a hand controller as standard equipment. This doesn't have a plethora of buttons and lights like those found on the more sophisticated scopes, but it works well at adjusting the scope's speed by the small amounts necessary for photography or image centering at high power. If an optional declination motor is purchased for the LX-10, the controller's north and south buttons can be used to adjust the scope in declination at a similar speed. The telescope can be used in either the northern or southern hemisphere, with this mode of operation being switch-selectable.

One thing I *don't* like about the LX-10's drive unit is the location of the control panel. The on/off switch; plugs for the hand paddle, focus motor and declination motor; and the red power on LED are mounted vertically on the drive base. With the scope mounted on its wedge, this panel becomes very hard to see. Even the illuminated power light is not obvious.

I like the LX-10. In fact, it has been the only telescope I've used with my university astronomy classes for the last several years. Even jaded as I am by 25 years of SCT usage, I wouldn't mind having an LX-10 of my own. It's just a very competent CAT. The optics are good, and the mount is quite usable, even for beginning imaging. How would it fare with novice astronomers? Most of my students have never seen a "real" telescope before they're introduced to the LX-10, and most are a little apprehensive about learning to use a "complicated

scientific instrument." After an hour or so of instruction, though, most are having a ball with this scope.
And by the end of the evening, they are positively
raving about "their" telescopes. If an SCT could speak,
this one would be saying. "Come on, I won't bite! Let's
have some fun looking at the sky!"

Meade has recently announced the availability of a
Deluxe LX-10. This is exactly the same OTA and fork-
mount as the basic model, but the small 30 millimeter
finder has been replaced by a much better 50 millimeter
unit, and the optional electric declination motor is
included as standard equipment. A copy of Meade's
Epoch 2000 Jr. computer star atlas software (a CD ROM)
is included in the package. At a price of about US$150
more than the basic scope, this is a decent bargain. The
larger finder, especially, would be a real help for the
average observer. The original LX-10 will, according to
Meade, also remain available.

Celestron Celestar 8

Meade's LX-10 is not the only game in town for the
budget-conscious or beginning SCT shopper. Celestron
has the Celestar 8. Like the LX-10, this telescope is
furnished with enhanced coatings on its mirrors and
corrector plate, which Celestron calls Starbright Coatings. Also like the Meade, the scope comes without
a case. But this telescope does come with a tripod
and wedge: actually, a tripod with a *built-in wedge*,
Celestron's "wedgepod." This really has very little effect
on the price of this telescope, though, which is about
the same as that of the LX-10, when the Meade tripod is
figured into the LX-10's final cost. Like all other
Celestron Schmidt–Cassegrains, the Celestar is available
with f/10 optics only.

If the Meade LX-10 was pleasantly light in weight, the
Celestar is *amazingly* easy to lift, even when mounted
on the wedgepod. It's significantly lighter than the
competing model. Unfortunately, I was not able to
carry the telescope around in its case as a test of
portability. Like Meade, Celestron has discontinued
cases as standard items for its CATs. Celestron does not
offer a low-cost option for the telescope comparable to
the cordura case from Meade. Oh, there are cases
available for the Celestar, but they cost several hundred
dollars, more than the average entry-level CAT fancier

Figure 4.3. The basic Celestar 8 (photo courtesy of Celestron International).

is likely to want to pay at first. I've always thought that the case was an integral part of the SCT system, very critical to making these scopes truly transportable. Without a real case, the telescope and its accessories tend to spread out all over your car or living room.

Once I'd carried the Celestron out into the field (this amazingly low weight telescope could actually be picked up mounted on its tripod and moved outside with no trouble at all), I quickly became very fond of this little CAT. The OTA was up to Celestron's usual high standards, with notably smooth and easy action of the focuser (though some observers do prefer the slightly stiffer action found on the Meades). The fit and finish of the entire telescope were simply outstanding – just as good as what I've seen on the higher-priced Celestron instruments. The Celestar's black gloss coat with orange lettering makes for a very attractive, high-tech look. One recent addition to the OTA of this and all other Celestron telescopes is the Faststar secondary mirror assembly. With this system, the SCT's secondary mirror can be removed, and a special CCD camera and optics assembly mounted in its place for wide-field f/2 imaging.

Performance? The Celestron OTA was no slouch. Images in a carefully collimated Celestar I tested were very good, being as nice as those I'd expect from *any* 8 inch telescope of any design. They were easily as good as those found in the Meade LX-10. What about the mounting? The Celestar is about as lightly built a Schmidt–Cassegrain as I've ever seen. But, like the Meade, this lightweight design didn't seem to hurt the telescope's steadiness very much at all. My "rap test" (strike the OTA sharply while observing at about 200× and time how long it takes the vibrations to die out) came out with results very similar to what I found on the LX-10, about 3–4 seconds for the shakes to die out completely. As with the previous scope, I judged this to mean that the scope's mounting is acceptable, if not rock solid. I attributed at least a part of the scope's vibrations to its undersized tripod and wedge.

The Celestar's drive also passed my inspection with flying colors. Its power source is a very convenient pair of 9 volt transistor radio batteries. Celestron advertises that this will power the telescope's stepper motor drive for up to 50 hours of use. I found this to be a fairly accurate figure assuming no optional items like a declination motor were attached and that the CAT was not used in cold weather (which will radically shorten the life of these little batteries). At first I was a little disappointed to find that the Celestar uses a spur gear drive instead of a worm set. But I didn't remain disappointed for long. The telescope tracked very well, and I didn't notice any practical difference between this CAT and its worm gear equipped competitor.

I did *not* like this telescope's included wedgepod. It was OK for general visual use, but for photography it would cause problems. Why? Because there is no way to point the wedge finely in azimuth to achieve a good polar alignment. The tripod legs are attached directly to the wedge, and the only way I found to move the unit left or right was by picking up the whole thing and turning it bodily. I was reduced to nudging (or kicking) the tripod legs in an attempt to get the wedge pointed at the azimuth of Polaris. This would only become an issue for someone interested in dabbling with through-the-scope photography, of course. You could certainly replace the wedgepod with one of Celestron's standard wedge–tripod combinations if picture-taking were your main goal.

Like the LX-10 tripod, this one is not adjustable in height and its legs are not held apart with a spreader. Also as in the LX-10, the Celestar I used was equipped with a barely adequate 30 millimeter finder telescope. This unit and its bracket are easy to remove, so an observer wanting to equip the telescope with a better 50 millimeter unit wouldn't have much trouble doing so. I would strongly advise that this be one of the first extras the new user purchases.

Like most SCTs these days – entry-level or advanced – the supplied accessories are kept to a minimum in the Celestar. Searching around I located the only provided items, a 25 millimeter Celestron SMA (super modified achromat) eyepiece and an inexpensive star diagonal. The SMA turned out to actually be a variation on the common and simple Kellner design of eyepiece. At this long focal length (of both the scope and eyepiece), the SMA worked well, but clearly was not quite as good as a higher-priced Plössl eyepiece such as that supplied with the competing Meade telescope. The stars at the edge of the field didn't look as good as they did in the Plössl, and the apparent field of view was decidedly smaller.

But don't get me wrong about the Celestar. The night I evaluated Celestron's Celestar 8 Hercules was riding high in the sky. Naturally, I turned the telescope to the Warrior's treasure, the great globular star cluster M13. The image of M13 in this telescope was, in my opinion, as good as I've *ever* seen in an 8 inch SCT. I lingered over the great ball of stars for *quite* some time (despite the owner's pleas for "just one quick look"). By the end of the evening it was clear to me that the Celestar is an excellent performing and very cost effective choice for the novice or economy minded amateur astronomer.

Meade 203SC/500 and 203SC/300 8 Inch SCTs

These two CATs, the 203SC/500 and the 203SC/300 (see Figure 4.4) are listed together here because they are basically very similar instruments – f/10 optical tube assemblies on medium and light-weight German equatorial mounts respectively. In price as well as capabilities, these SCTs are squarely in the budget/ entry-level camp and offer a minimum of features and

Figure 4.4. Meade's 203SC GEM SCT (photo courtesy of Meade Instruments Corporation).

included accessories. Both telescopes are equipped with Meade's Super Multicoated Optics group in a standard Meade OTA. The 300 is mounted on the LXD-300 GEM, while the 500 comes on the heavier-weight LXD-500 mount.

These are very attractive scopes. The naturally high-tech looking German mounts and the brushed aluminum tripods are complemented by the OTAs, which are finished in a striking white high-gloss paint rather than the usual dark Meade blue. Nice looking, and nicely priced at under US$900 for the 500, and not quite US$800 for the 300 at the time this is being written. The

OTAs on both of these scopes perform as well as any Meade 8 inch f/10 SCT; that is, very well. But there the similarity ends. The different mounts mean these scopes are actually very different animals when it comes to usefulness.

The 203SC/500 comes equipped with the usual accoutrements expected for an entry-level SCT, a 1.25 inch star diagonal and a single low-power 26 millimeter focal length eyepiece. Like the LX-10, this scope features a surprisingly good Meade Series 4000 Plössl eyepiece. As we've also unfortunately come to expect with the budget CATs, the 500 sports the typical too-small 30 millimeter finder. The telescope mounting is imported from Taiwan by Meade and uses good-quality bronze worm gears on both axes. This medium-weight GEM has slow-motion controls for both declination and right ascension which will allow users to track celestial objects by hand power. But like the other German equatorials in this class being offered by Meade and Celestron, motor drives are an extra cost add-on. The telescope's tripod features an accessory tray for eyepieces and a bubble level to help with setup and alignment. The mount can be equipped with a polar axis finder to aid in polar alignment, but this is, like the motor drives, an optional accessory.

How well can the 203SC/500 be expected to perform? Fairly well in my judgment – for visual work, anyway. The mount is reasonably steady, if hardly rock solid, with the SCT OTA mounted on it. The mount that I examined had rather stiff motions in both declination and right ascension. This was not a fatal flaw, but the telescope didn't move as easily as I would have liked. The problem seemed more the result of the heavy grease used in the mount, though, than any design flaw.

I attributed much of the mount's vibration to the tripod, just as I did with the entry-level fork-mount telescopes' tripods. This tripod is substantially lighter even than the one found on the fork scopes, and seems much shakier and more flexible. The tripod is the Achilles heel of the LXD-500 mount and is close to being what I'd call substandard for use with an 8 inch telescope.

Meade offers two drive options for the LXD 500 mount, a dual-axis clock drive and a single motor model. The model 1702 dual-axis unit features motors for both the RA and declination axes, and runs at drive rates from $2\times$ sidereal speed, which is good for photo guiding, to 32×, which can be used for moving the telescope short distances across the sky and for centering objects. The model 1701 single-axis drive system is the same as the dual motor model as far as speeds and power source, but it includes only one motor. The single-axis system cannot be upgraded for dual-axis operation.

What's the final verdict on the 500 SCT? Its strong points are its fairly low weight (49 pounds for OTA, tripod and mount) and its low price. It would be practical to pick the whole scope up and quickly trot outside for a quick look as long as stairs and long distances would not have to be negotiated. Removing the OTA from the mount is very easy, and this would make the scope extremely easy to carry into the backyard in two trips. The tripod collapses very quickly by removing one knob from the accessory tray (usually this style of tripod is a pain to fold up), making the mount even more portable. Once set up, the scope is reasonably competent, if a bit shaky. While some photography could be attempted with this telescope, it is really best suited to visual use.

What about the 500's sister telescope, the 300? I was frankly amazed by the price of this SCT. I don't believe there's *ever* been a complete, working CAT offered from either of the manufacturers for a price this low. Less than US$800 in 1999 is simply unreal! How do they do it? They do it by replacing the 500's medium GEM with another Taiwanese import that can only be described as "lightweight" in every sense of the word (in real terms, the entire 300 SCT is only about 5 pounds lighter than the 500, though). The 300 features the same tripod setup as the 500, but the equatorial head is noticeably smaller. It is also more cheaply made, and uses aluminum worm gear sets on both axes rather than the more precise and hardy bronze gears of the larger mount. It does feature most of the same niceties as the 500, though, including dual slow-motion controls, and a provision for the installation of an optional polar alignment borescope.

How well can this small mount work? It certainly looks dwarfed by the 8 inch OTA in the pictures. This scope/mount combination does work, and the telescope can provide some pleasing views at low and medium powers. But there is no escaping the fact that this is about as inexpensive and downsized as you can get before the whole house of cards collapses. The LXD 500's sometimes annoying tendency to vibrate becomes an almost constant irritation with this telescope. In a breeze, or when focusing, or when using the slow-motion controls, this scope's case of the "shakes" becomes all too obvious at any but very moderate magnifications. There are a number of things the 300 owner could do to help steady this telescope, but a scope of this size on a mount this light will never really provide for steady viewing.

Celestron G8

Meade is not alone in offering an 8 inch GEM Schmidt–Cassegrain to the amateur astronomy market; Celestron has a long history of packaging its OTAs with medium-duty German equatorials. This started in the 1970s with the famous Celestron Super Polaris C8, which featured a nice GEM made by the Japanese telescope maker, Vixen. Not long ago, Celestron switched to Vixen's updated version of this mount, the Great Polaris, and very recently has begun shipping the telescope with a similar but non-Vixen made mounting. With this

Figure 4.5. Celestron's GEM scope, the G8 (photo courtesy of Celestron International).

change, the telescope has been rechristened the G8 (see Figure 4.5).

The new Celestron G8 consists of a standard f/10 Celestron optical tube with enhanced coatings and the Faststar-compatible secondary. This familiar OTA is mounted on the new equatorial head and tripod that the company is importing from China. This mount, which many amateurs are referring to as a clone of the Great Polaris, has been dubbed the CG5. Included with the G8 are an adjustable aluminum tripod with accessory tray, a 90-degree star diagonal, and a 25 millimeter focal length SMA super modified achromat eyepiece.

Is the G8 a worthy successor to the Celestron Super Polaris of days gone by? In my opinion it is. The G5 mount is not oversized, but it is adequate for this introductory telescope's job – general visual observing. Many amateurs remember the Celestron Super Polaris GEM scope fondly, but, frankly, its mounting was not exactly the Rock of Gibraltar, either. In many ways, the G5 compares quite favorably with the Vixen units. What is the most glaring difference between this mounting and the Vixens? What I noticed was the quality of assembly and manufacture, and not so much the weight or durability.

This mount is a little "sloppier" than its Japanese sisters, exhibiting more backlash and similar problems. The parts are just not made to very exacting tolerances. And the way the parts are produced is also different. From what I can tell, all the components of the equatorial head are cast rather than machined. Some of the assemblies that are solid pieces of metal in the Vixen Great Polaris and Super Polaris mounts are hollow in this new GEM design. A number of recent purchasers of this telescope have also complained that the grease used to lubricate the mount is really more like *glue* than grease, and serves to make this equatorial work less well than it really should. Grease can be cleaned off and replaced with a lighter lubricant. Actually, this glue-grease seems to be the source of most of the reported problems with this mount, which usually involve stiff motions in both axes and motor problems that can be traced to this stiffness. With a disassembly and relube and a little filing down of rough surfaces, this mount can perform surprisingly well.

Feature-wise, this is another staunchly entry-level CAT. We have the frustratingly small 30 millimeter finder, the inexpensive eyepiece and the light, bare-bones, driveless mounting. The G5, like all of these imported telescope mountings, is equipped with setting circles which are just a little bit too small to offer much in the way of precision. One feature of the telescope I like a lot is the nice dovetail-style sliding bracket that attaches the telescope tube to the G5 mount. This simple, yet elegant arrangement makes it easy to get the OTA on and off the mount.

The G8's mounting, like others in this class, is shaky, but not outrageously so. I would not hesitate to classify it as "usable" for at least low and medium power observing. Once again, a poorly designed and light-weight aluminum tripod is the main culprit. This mount would be a good deal steadier if only it were equipped with a slightly more stable tripod. Celestron offers both a dual axis drive and a simple right ascension motor. One nice thing about the mounting of the G8 is that any drive system sold for Vixen's GP mountings can be used on this GEM without modification. This includes Vixen's amazing Skysensor 2000 PC goto drive system.

Who would like the G8? Anybody needing an inexpensive but usable Schmidt–Cassegrain. The tele-scope, like Meade's similar models, is *not* perfect, but it could easily provide a lot of viewing pleasure. I would advise anyone considering the purchase of this telescope, though, to be prepared to tear down this mount, remove the glue-like grease and relubricate it. If you don't feel up to this task, it's likely that one of your area's amateurs can help you. Any person good with tools and mechanical equipment should be able to assist you in this necessary task.

The Mid-Priced CATs

The telescope models listed below fall into the middle ground between the entry-level SCTs and the most expensive CATs. Their performance can be quite impressive. In fact, just 10 years ago, before goto SCTs, these two would have probably been consid-ered "flagship" telescopes for both companies. If you're looking for a well-made CAT able to take on the challenges of astronomical imaging, these two definitely fill the bill.

Figure 4.6. Meade's mid-priced LX-50 SCT (photo courtesy of Meade Instruments Corporation).

The Meade LX-50

Like the rest of Meade's current crop of Schmidt–Cassegrains, the attractive blue tube of the LX-50 (see Figure 4.6) houses Meade's enhanced optics group. The telescope's hand controller is a standard item, as is a 25 foot (7.7 meter) DC power cord for operation from an external battery source. The Meade standard wedge is also included in this package, but, as is the case with the LX-10, the tripod is an extra-cost option. A tripod must be purchased immediately, of course.

The LX-50 8 inch SCT currently retails for about half as much again as the budget-priced LX-10. What do you get for your money? Is the extra amount worth it? What's the difference? One look at this scope and you'll recognize that this is an obviously more "serious" instrument. Both the fork-mount and drive base are much heavier than those of either the LX-10 or Celestron's Celestar. I wouldn't necessarily describe the fork on this telescope as "massive," but it has been significantly strengthened. Differences are also apparent when the hand controller is examined. In addition to the directional buttons that allow north, south, east and west guiding corrections at 2× sidereal speed, there is also a speed selector button that allows users to increase the drive's rate to 8, 16, and 32 times sidereal speed. A rate of 32 times sidereal is still not fast enough to slew the telescope from one part of the sky to the other, but it is very useful for moving a few degrees and will greatly aid in rough centering of objects. In addition to this more sophisticated right ascension drive, the Meade is fitted with a built-in declination motor, making this telescope nearly photography ready out of the box. Pressing the north/south buttons on the hand paddle activates this motor, which can be driven at the same 2/8/16/32× sidereal speeds as the RA drive.

The LX-50 sounds like quite an advance over the humble entry-level CATs, but how much better does it *perform?* In my judgment, substantially better, especially if you're trying to take pictures. The heavier fork is a big help here. The beefier mounting of the 50 makes photography and high-power observing much more practical in breezy conditions. On all but the calmest nights, I felt photography would be an exercise in futility with the LX-10. The LX-50's heavier fork and drive base also mean that accidentally brushing or bumping into the scope during a photographic exposure wouldn't necessarily ruin a picture. This more substantial fork-mounting does result in a heavier telescope. LX-50, wedge and tripod come in at 71 pounds compared to 49 for the complete LX-10.

During my tests of the LX-50 I enjoyed the versatility of its multispeed drive, though I didn't find myself using the higher drive rates very often. This is probably just me, though. I've used SCTs with more primitive drives for so many years, that I found myself automatically reaching for the LX-50's normal mechanical RA and dec slow-motion controls rather than the hand controller buttons. The manual slow-motion controls worked smoothly, but I'm sure that after a few months of using an LX-50 I'd get more used to the luxury of the motorized slomo options. The basic RA drive on this telescope works well and seems completely sufficient for photography. There was a fair amount of periodic error present in the telescope I evaluated, but it was smooth and regular. The error would be easy to guide out during a long exposure and not a hindrance to deep sky imaging.

In keeping with its more sophisticated drive and electronics, the LX-50's controls, which are conveniently located on a slanted panel extending out from the drive base, are much more elaborate than those examples found on less expensive CATs. In addition to an on/off switch and red LED power indicator, the panel has inputs for the built-in dec motor's connector, a CCD autoguider (for photography), an external battery cable, and a hand controller. An auxiliary connector is also present to power devices like illuminated cross-hair eyepieces from the scope's power supply. The compartment that holds the scope's six AA batteries is easily accessed by removing a plastic cover on the control panel.

The RA drive on the LX-50 works well, but it is somewhat surprising to find an SCT in this price class that lacks PEC. PEC, *periodic error correction*, is a feature developed by SCT manufacturers to improve the accuracy of their telescope RA drives for photography. The concept is simple. The polar-aligned scope is pointed at a bright star, the drive is turned on and a button is pushed enabling the drive's PEC circuitry. A cross-hair eyepiece is then placed in the star diagonal. When the star wanders out of the cross-hairs, the observer pushes the appropriate button on the hand controller to move the star back into the center. This is exactly what would be done if a photograph were being taken. But with the PEC enabled, something special is happening. As the observer continues to guide with the hand paddle, the telescope drive "remembers" the button pushes that are made. After a specified amount of time the user stops guiding and turns off PEC. When a real photograph is to be taken, PEC playback can be turned on. The scope adjusts the drive just as the observer did when the PEC recording was made. In theory, this should go a long way to eliminate manual guiding of the telescope during picture taking.

But the LX-50 *doesn't* have PEC. Does this mean you can't take good astrophotographs with this otherwise capable CAT? Not at all. Meade and Celestron have both pushed the PEC concept over the last decade, and this has given some aspiring photographers the idea that deep sky images can't be done *without it*. It *is* a big labor-saver for the imager, but it is *not* necessary. SCT astrophotographers were taking wonderful pictures long before PEC was invented. Using a non-PEC drive only means that the photographer has to pay closer attention to guiding. I wish the LX-50 drive had PEC, but its lack doesn't seriously compromise this CAT's usability.

Accessory-wise, the LX-50 is fairly well appointed. The new user will find a good-quality star diagonal and the nice Series 4000 26 millimeter Plössl in the box. This SCT, just like the budget gang, has a small 30 millimeter finder mounted to its OTA. This cries out for quick replacement with a larger model.

Just as this book was being completed, Meade announced that the LX-50 series of telescopes is being discontinued. The 50 will remain on sale until existing stock is depleted, which may be some time. But many industry watchers believe that Meade plans to replace the venerable "50" with a similar mid-range telescope featuring computerized goto pointing.

Celestar 8 Deluxe

Except for the name Celestar, this telescope (see Figure 4.7) has little in common with Celestron's basic Celestar entry-level scope. Scanning the dealers' price lists, you'll find that the Celestar Deluxe costs almost twice as much as its similarly named sister scope. Why? Because this mid-range CAT is a significant advance over the basic Celestar in every way. The fork-mounting is similar to that on the entry-level scope, but the drive base is noticeably bigger and stronger looking. Although Celestron's designers beefed up the Celestar Deluxe quite a bit, it's still a pleasantly portable CAT, weighing in at 52 pounds as compared to the 41 pound weight of the Celestar basic. But the drive base is not the only difference between these Celstrons.

Figure 4.7. Celestron's Celestar Deluxe (photo courtesy of Celestron International).

The most important advance in the Deluxe comes in the electronic drive system. Gone is the spur gear RA system of the basic, replaced by a high precision worm gear. This drive features four tracking rates, *solar*, *sidereal*, *lunar*, and *King*. It can also be slewed at 8× sidereal rate as an aid to centering objects. The Deluxe can be moved in declination by using the manual slow-motion control, but it is also furnished with an electric declination motor. The included hand controller allows guiding and slewing at 8× in both axes via the four directional buttons. Speed selection and other functions are handled by buttons on this SCT's well-appointed front panel. The drive also features the desirable PEC recording feature.

The control buttons, indicators and input plugs for this scope are mounted horizontally on a panel extending from the drive base. Plugs are provided for the declination motor, 12 volt external power input from the furnished "cigarette lighter" cord, and input from the hand controller. In addition to an on/off switch, the panel features buttons for drive rate selection and PEC operation. LEDs indicate drive speed and PEC mode. The periodic error correction circuit for the RA motor allows users to record guiding corrections, making the telescope's excellent worm gear drive even more accurate.

Unlike the basic Celestar, this telescope is equipped with the much better and more versatile standard Celestron wedge and tripod. The basic's aggravating wedgepod has been replaced by a traditional tripod/ wedge combo that performs well for both visual use and photography. A serious photographer would, however, be advised to purchase Celestron's optional fine attitude adjuster kit for the wedge. The components included in this kit provide the wedge with altitude and azimuth fine adjusters, enabling the very precise polar alignment required in photography. Another plus for imagers is the telescope's Faststar optics, which allow f/ 2 imaging with compatible CCD cameras.

On the observing field, the Celestar Deluxe is a very pleasant and capable CAT. It's almost light enough to be carried from the house to the backyard in one piece – though it should probably be removed from its tripod for safe transport. Once set up and operating it is truly a joy to use. The drive tracks very, very well. The extra drive rates, lunar, solar, and King, worked as advertised, and seemed accurate. The King rate, by the way, is a special drive speed that takes into account the effect refraction has on the images of bodies near the horizon. This might come in handy for some photographers, but most imagers studiously avoid taking pictures of objects anywhere near the horizon due to "seeing" turbulence.

The Celestar Deluxe does a fine job with cameras and CCD imagers, but it might also be considered if visual observing is the observer's only goal. The strong yet light fork-mount provides steady views at high power, and the Celestar I tried had some of the best optics I've seen in an 8 inch f/10 Schmidt–Cassegrain telescope. The smooth drive and good selection of accessories make for a very rewarding CAT experience.

The Deluxe's accessories proved to be about as good as I could have wished. A nice 50 millimeter finder is standard and really helps in chasing down objects in bright suburban skies. Instead of the so-so SMA eyepiece that comes with the basic scope, the new Deluxe owner finds a good quality 26 millimeter Celestron Plössl. The star diagonal also seemed to be of better quality than the one used on the less expensive telescope.

This CAT impressed me mightily; it would be difficult not to recommend it. The Celestar Deluxe model is a perfect match for an amateur interested in celestial picture taking. If you want to pursue the difficult art of astrophotography and have no interest in computerized goto drives, this could well be your best choice in a new telescope.

Celestron sells both the Celestar Deluxe and the Celestar basic in computerized versions. All this means, though, is that the manufacturer ships the telescopes with the Celestron Advanced Astro Master digital setting circle computer system. It is not even installed – the purchaser must do that simple job. This computer provides readouts only; it does not control the telescopes' drives in any way.

Luxury CATs

These are the CATs that, because of their prices and features, can be described as "luxury CATs." They are the pinnacle of the current outputs of Meade and Celestron.

Meade LX-200 8 Inch

The LX-200 (see Figure 4.8) was a revolutionary telescope when it was introduced in 1992, but when looking at the initial advertisements this was easy to miss. The picture of the scope was impressive, showing a shiny blue-tubed Meade OTA on a fork-mount that was obviously heavier-duty than the company's earlier models. But it didn't really look like much of an advance over the Premiere, the premium series of SCTs Meade had been selling up until this time. The ad copy didn't reveal much either, with most of the column space touting the telescope's "most rigid" fork-mount and the "four speed operation" of the LX-200's drive. Not until the eighth paragraph are some words about the new telescope's goto operation. Buried in this text was the remarkable claim that the telescope would *automatically slew to any one of 747 objects* (this seemed like a huge number at the time).

When amateurs realized Meade was embarking on a program of selling telescopes that pointed themselves, they were intrigued. Seemed like it might be a good idea. But could it possibly work? Didn't seem likely. Computers were still an unknown quantity to many

Figure 4.8. Meade's legendary LX-200 (photo courtesy of Meade Instruments Corporation).

even fairly technologically oriented people at the time, and the few that had been used in astronomy hardware hadn't worked very well at all. Would the LX-200 be any different?

Very different as it turned out. From the day of its introduction, the LX-200 proved the worth of computerized pointing. It was this single telescope model that forever changed the way we look at SCTs. The 200 has aged gracefully, going through a number of changes, mainly in the software realm, which have served to keep it up-to-date. The LX-200 works and works well, and has undoubtedly opened the universe to many people who'd never have been attracted to the hobby before.

The LX-200 is an expensive telescope, at least in the reckoning of most amateur astronomers. The price for the 8 inch model is roughly double that of the LX-50. What does such a huge premium get the CAT user? Normally, the usual stock Meade f/10 OTA, the optics of which have definitely proved their worth. If you're a photographer or are just interested in wide-field views, you also have the option of ordering the scope with Meade's f/6.3 optics. The LX-200 also features a good, heavy-duty fork-mount that, with the exception of a few minor differences, looks identical to what's found on the LX-50. In fact, the whole thing looks very much like the 50.

But it's what's inside that counts. This telescope's drive base contains a computer motherboard rather than just the simple drive corrector circuitry and DC motor of the LX-50. Electronics are also packed into the fork, which features a built in worm-gear declination drive and control circuitry. Perched on its included tripod the scope looks high-tech, but not overly imposing. The first real clue that something is different about this CAT comes when the new owner takes a look at the furnished hand controller.

The SCTs reviewed thus far have used hand controllers that are adorned with a few simple buttons: four directional switches, maybe a couple of extra buttons for speed selection. These simple remote controls don't really prepare the owner for the LX-200 controller, which is the nerve center for all of the telescope's operations. Instead of a few keys, this paddle is equipped with a pressure-sensitive membrane-style keyboard and an LCD display. Close examination shows that there are 19 different keys, and that many of them have more than one function. Not only does this hand controller *look* like a little computer, it actually *is* one. Inside the unit is a dedicated 68CH05 microprocessor and memory.

The controller, which looks a lot like a fancy cell phone (or a Star Trek communicator) plugs into a front panel that is also more complex than what is usually seen on SCTs. In addition to the hand controller port, there are inputs and outputs for CCD guiders, an external computer via a standard RS-232 interface, an illuminated reticle eyepiece, a motorized focuser, input power and a connector for the telescope's declination drive system. There are switches to turn the scope on and off and to set the drive motor's rotation for the northern or southern hemisphere. There's also an LED bar-graph-style ammeter to indicate the current being drawn by the telescope.

I'll fully admit that the first time I was confronted with a 200 with its seemingly endless array of switches, dials, lights and computer menus, I felt completely overwhelmed. Sure, I could polar align a telescope with the best of them. But get a computer to track the stars? I didn't know if I was up to the task or not. Luckily, getting the LX-200 operating is very simple. The telescope has numerous options and capabilities, but I discovered that setting up for simple tracking and goto pointing is very easy. All that's really required for basic operation is for the user to input date, time and location into the scope's computer via the hand controller. The nice red-lit LCD display makes everything surprisingly simple. The user does have to take care to get the right time zone set and time entered as accurately as possible. Once the telescope knows *where* it is and *when* it is, all that remains is for the user to align the telescope on a single star. When the single star alignment is complete the LX-200 will locate and track objects in the alt-azimuth mode.

One of the most surprising sights for the Schmidt–Cassegrain veteran is seeing an LX-200 working in its *normal* alt-azimuth mode. The scope drive base is bolted directly to the tripod – no wedge is used. Because it is blessed with dual motors and a very smart computer (the LX-200 uses the Motorola 68000 microprocessor, the same one used in many of the earlier Apple Macintosh computers), this scope doesn't need polar alignment. It will happily track any object across the sky as long as location and time have been correctly entered and the user has centered that one star from the long list of alignment stars.

There are some very good reasons for mounting a CAT in this fashion. One, of course, is that you don't have to pay for a wedge (no wedge is included with the LX-200 scopes), helping Meade to keep the price of this computer-packed marvel relatively low. Another plus for wedgeless operation is enhanced stability. Tipping the fork and drive base of an SCT over at an angle to point at the celestial pole makes for an off-balance condition. In alt-azimuth mode the fork and drive base ride squarely on the tripod head. This makes a real difference. The LX-50 and the 200 use what is essentially the same fork-mount. But the LX-200 is *much* more stable than the polar aligned

LX-50 due to the alt-az arrangement. If an LX-200 user wants to attempt astrophotography, the best option is to purchase an optional wedge for the telescope – either Meade's standard wedge or its more stable super wedge.

How well does the LX-200's complex machinery perform out in the middle of a dew-heavy and cold observing field? Just as advertised. The computer drive systems on the LX-200s I've tried have worked well and consistently, usually placing any one of the telescope's huge library of objects in the field of view of an eyepiece or a camera. There is certainly no lack of celestial sights to choose among. That seemingly huge database of 747 objects in the original LX-200's database has, over the years, been increased to 64,359! Many of these will be invisible in an 8 inch (or even a 10, 12, or 16 inch) LX-200, but they will be well within the range of film and CCD cameras.

The LX-200 is obviously a dream telescope. Why would anybody *not* want one? As was mentioned earlier, computerized telescopes will *never* be as reliable as simple models whose most complex component is a battery-powered motor. Electronic parts can and will fail in *all* computers – on the desktop or in the telescope. To its credit, Meade has a good support system in place for these telescopes – they can be fixed in a timely fashion without undue trouble. But the fact remains, if you choose a highly complex instrument like this it's a good bet it will be out of action for repairs occasionally. Due to the infamous Murphy's law, expect that this will happen at the most inopportune moments, too – eclipses, planetary apparitions, and the passages of comets seem to be favorite times for my astro gear to glitch! Don't let this scare you off, though. The LX-200 is a mature product and is probably as trouble free as possible, considering the fact that the price must be kept reasonable. Desktop computers can and do break down, but that doesn't stop most of us from using them every day.

Celestron Ultima 2000 8 Inch SCT

The LX-200 made a huge splash when it was introduced in 1992, and the ripples have been spreading ever since. This Meade telescope quickly became the most wanted

Schmidt–Cassegrain since the introduction of the original C8. It was clear that Celestron would have to respond with a goto instrument of its own if it wanted to remain competitive in the world of high-tech CATs, and there were rumors that Celestron was preparing to introduce a "robo-scope." But several years passed, and there was still no "Celestron LX-200." This seemed strange. After all, what was so hard about doing a computerized telescope? Celestron had even produced a workable, if slow and problematical goto SCT, the Compustar, in the mid-1980s, years before the LX-200 was born. When Celestron finally brought a new goto telescope, dubbed the Ultima 2000, to market in the mid-1990s, it was immediately clear what had taken so long. Celestron had chosen to redesign their goto scope from the ground up.

The Ultima 2000 (see Figure 4.9) is a striking looking telescope. The light gray of the tube is a color that's different from any Celestron has used before. Aside from the standard design Faststar-capable OTA, everything else is different too. At first glance, the light fork is similar to the units used on the Celestar telescopes, but the long-time SCT user is in for a real surprise when the mount is examined a little more closely. Where are the manual slow-motion controls and the declination and right ascension locks? There are none.

In many ways, this SCT is quite similar to the Meade LX-200; it's a goto scope normally operated without a wedge in alt-azimuth mode. But the design of the Ultima is different, and using one is a very different experience. There are no locks or manual slow-motion controls on the scope because Celestron has designed this SCT around clutches on the declination and right ascension axes. Want to move the telescope by hand? Just grab it and move it. No locks to unlock. Nor is there the problem of manual movement making the telescope "forget" where it's pointed. The Ultima uses optical encoders that, unlike the LX-200's units, are separate from the drive motors. Once aligned, the computer continues to receive information about the scope's attitude whether it is moved by the motors or by hand. The Ultima 8 drive system is surprisingly different as well.

Like the LX-200, the Ultima 2000 features worm gear drives on both the declination and right ascension axes. But the Ultima sports two motors for each axis, a slow motor and a fast motor. According to Celestron, this improves reliability and helps keep down the noise level

Figure 4.9. Celestron's goto CAT, the Ultima 2000 (photo courtesy of Celestron International).

during slewing (the average LX-200 sounds a little like an overworked coffee grinder when it is moving at high speed). When the command is given to point the telescope at an object, the Ultima starts out in its quick 15 degree per second rate, driven by the fast motor. When the target is approached, the scope automatically switches to the slow motor for fine pointing. The Ultima's motors can also be run manually from the included hand controller. Eight speeds from a 2× sidereal guiding rate to the high-speed 15 degree per second mode are available from this controller.

The Ultima hand controller also diverges from the Meade example. The Ultima controller is surprising for the simplicity of its design. There are four directional control buttons, four function buttons, and an LED display. That's it. All computer functions are controlled by the four buttons that reside directly beneath the LED: *menu, enter, up,* and *down.* The Ultima hand controller works well, but it really reminds me more of one of the popular digital setting circle computers in use by amateurs than a real hand-held computer. Like the controllers for electronic setting circles, this device cleverly makes use of a few switches to select menus, enter data and make choices. Many beginners will find the Ultima hand controller initially much less intimidating than the more complex-looking LX-200 unit.

Like the hand controller, the Ultima's control panel seems very bare, even in comparison to some non-computerized telescopes. There's an input for the hand controller, for external power and for a CCD autoguider. There's also a power switch, and that's all. The computer contained within the drive base is complex and very capable, though. Its database contains a library of 10,000 objects, more than most amateurs will ever need. For really advanced operation, the Ultima 2000 can be controlled by an external computer running many of today's popular astronomy programs.

Accessories are about what you'd expect for a telescope of this caliber and price (it presently costs slightly more than the Meade 8 inch LX-200). In addition to a good star diagonal, the telescope comes with a 26 millimeter Plössl eyepiece and a 9 × 50 finder scope. The standard Ultima 2000 package includes a field tripod, but no wedge. I was dismayed to learn that Celestron has recently discontinued the hard case that had formerly been included with the Ultima 2000. This nice case made the already light Ultima 2000 very convenient and transportable and will be missed.

In its abilities, if not its design and layout, the Ultima 2000 is similar to the Meade SCT. But does it perform as well as the LX-200 on the field? I think so. The telescope is a delight to set up. At 28 pounds, the Ultima in its hard-sided case is easy to carry out of doors. Positioning the telescope on its sturdy tripod is quick and simple. The Ultima can be placed on an optional wedge for photography, but it is usually mounted directly to the tripod for alt-azimuth operation. The light load of the Ultima doesn't stress out the tripod and makes for a very stable telescope, especially in the alt-az configuration. The U2K's alignment process is both easier and more complex than that of the LX-200. The user doesn't normally have to enter date, time and location; all that's required is just aiming the telescope at two alignment stars.

Once aligned the telescope tracks well and does a fine job of pointing to selected objects. The Ultima

2000 is not intimidating; all the gizmos work in a smooth and intuitive fashion. The ability to move the tube by hand without worrying about locks and gears and without losing my place in the sky was a definite attraction for me. The clutch arrangement seemed strange at first, but being able to move the telescope by hand seems much more natural than motor-only operation. Every time I've used an Ultima 2000, I've enjoyed the experience of zipping across the sky from object to object with ease. I've found myself observing more deep sky wonders in one night than I usually visit in a week of observing with a manual telescope. A goto scope is a guilty pleasure I could learn to really love.

There hasn't been a perfect SCT yet, and the Ultima 2000 is not an exception to this rule. I *did* find a few things to complain about. The lockless clutch-style operation is convenient, but it comes at a price. These clutches must be adjusted carefully by the user for just the right amount of tension. This is not overly difficult, but even when they are adjusted correctly, balance can become a problem. Counterweights for mounting on the tube and fork are included with the Ultima 2000, since proper balance is critical for the telescope's goto operation. If the scope tends to slip on its clutches, pointing operation will be adversely affected, often in a severe manner. I have not found this to be much of a problem when doing visual observing with standard 1.25 inch accessories, but the use of a 2 inch diagonal and heavy eyepieces, or mounting a heavy camera on the rear cell, can cause some balance headaches with this telescope.

The dual motors on each axis seem to help make the telescope much quieter than the LX-200, but they also make the telescope slower. You'd think that at a top speed of 15 degrees per second, the Ultima would put you on target a lot sooner than the 8 degrees per second Meade scope. In testing, I found this not to be the case. The LX-200 actually seems to point to selected objects *more quickly* than the Ultima does, largely due, I think, to the U2K's system of switching to the slow speed motors when it gets close to an object.

The Ultima's hand controller worked well, but for the occasional user, the data entry system could be a problem. After using the scope for a while, remembering the required button pushes and combinations became easy and intuitive. But after being away from the scope, I tended to forget how to use the controller and felt like I was starting back at square one. For me, using the more complicated looking Meade hand paddle is actually simpler.

Finally, all the reliability caveats mentioned for the LX-200 also apply here. This is a complicated telescope which, while well made, is more prone to failure than a simple Celestar. The Ultima 2000 appears to be a very reliable telescope, but not more reliable than the LX-200. This CAT, like its cousin, has a welter of electronic components, some of which *will* eventually fail.

Despite these reservations, I still like the Ultima 2000 very much. It is an innovative telescope, and not just a copy of its Meade predecessor. Its chief advantage is its lightweight build. This would be the perfect telescope for the amateur who wants a goto telescope primarily for visual use. This SCT, which can take photographs like any Schmidt–Cassegrain and use CCD cameras in the Faststar configuration, can be a very capable celestial imager. But a heavier mounting would be more desirable for traditional long-exposure film-type astrophotography. Lightweight though it is, the U2K nevertheless still has a solid, quality feel like its illustrious Celestron forebears.

Big CATs

When most amateur astronomers think "Schmidt–Cassegrain telescope," they also usually think "8 inches." This is the most common size of SCT; so common that it has truly become an icon for an amateur telescope. But from the beginning, manufacturers have produced SCTs for users who want, and can afford to pay for, larger aperture. Though commercial Schmidt-CATs have been made in sizes up to 22 inches, the largest production model currently available is a 16 inch.

Meade LX-50 10 Inch

The 10 inch LX-50 is very much a scaled up version of the 8 inch. It uses the exact same drive base, wedge and tripod. The fork is a little larger in order to accommodate the 10 inch OTA, but it is very similar to the smaller telescope's mount in strength of construction. The OTA held by the fork is available, like the 8 inch model, only in an f/10 focal length. There *is* one considerable difference – other than aperture – between the two models of LX-50, *weight*. The total (scope and tripod) for the 8 is a manageable 71 pounds. But add the extra 2 inches of aperture and this figure shoots up to 89 pounds. The 10 inch telescope's OTA, fork and drive base come in at a considerable 55 pounds. The user has to lift this 55 pounds onto a wedge every time the scope is assembled. This might not seem too bad in a well-lit living room on a quiet Saturday morning. But remember, this big fellow will have to be maneuvered off the wedge and tripod in the dark after a long evening of observing. One other consideration is that, because this scope uses essentially the same mounting as its little brother it will *never* be quite as stable as the smaller CAT.

Like the 8 inch LX-50, the 10 inch is being discontinued by Meade. It will remain on sale for a while longer, though, and represents an excellent bargain for the observer wanting a simple, no-frills, large-aperture CAT. Due to the large numbers of LX-50s that were produced, expect this telescope to be a staple on the used market for many years to come.

Meade LX-200 10 Inch

The LX-200 10 inch is also very similar to its smaller-aperture equivalent: same drive base, enlarged but similar fork, same tripod. Same choice of f/6.3 or f/10 optics. The weight increase for the larger telescope is similar to what we saw with the 8 and 10 inch LX-50s. The 8 inch LX-200 comes in at 69 pounds and big brother is 86 pounds total. The weight that has to be lifted onto the tripod of the 10 inch is 58 pounds as compared to 41 pounds for the 8 inch. Due to the lack of a wedge, the 10 inch LX-200 is actually lighter than the LX-50 10 inch when the entire scope is considered. Many people find this telescope much easier to set up than the LX-50 10 inch model. The reason for this is that the LX-200 normally doesn't have to go on a wedge and this makes the setup process much less difficult.

LX-200 12 Inch

Those LX-200s just keep getting bigger! If the 10 inch is big and a little heavy, the 12 inch can best be described as massive and a little backbreaking. The total weight goes to *120 pounds* with this telescope. The user will be lifting an OTA, fork and drive base combo of 70 pounds to the tripod. I *have* seen this done by one person, but a friend would be a big help at setup time! Steadiness-wise, the LX-200 12 inch keeps up with the smaller models because of its included *giant field tripod* (50 pounds). The telescope uses the same drive base and an upsized version of the same fork as the smaller LX-200s.

There are few changes in the 12 inch other than the weight increase. The corrector plate on this telescope is made of BK-7 glass, which is superior to the common float glass used in most Schmidt–Cassegrain corrector lenses. Because of the 12 inch OTA's heavier weight, the nine-speed LX-200 drive has been reduced to a seven speed unit for this model. The top speed is 6 degrees per second, somewhat slower than the 8 degrees per second of the smaller apertures. There is no f/6.3 12 inch. This optics set is only produced in an f/10 focal length.

Is the 12 inch a viable option for the average amateur? Perhaps. The US price, which is substantially higher than that of the 10 inch, will give some prospective owners pause. But the size and weight of the telescope will be the stumbling block for many more. Make no mistake, this is a *big* telescope. Even if there's a friend handy to help with setup, many amateurs may find this SCT less than practical for use as a primary instrument. Who wants to put together a 120 pound scope for a 20 minute look at the Moon on a Monday evening? Still, this big beast does exercise quite a bit of attraction for deep sky observers. It delivers over 30% more light than a 10 inch. Twelve inches seems to be the aperture where many brighter deep sky objects begin to look spectacular visually. The 12 inch LX-200 is an impressive instrument, but it is much more practical if it is installed in a permanent observatory.

LX-200 16 Inch

For the 16 inch, observatory mounting on a permanent pier isn't just recommended, it's almost required. This great big thing (see Figure 4.10) weighs a total of 313 pounds! The OTA alone is 120 pounds, not something to be casually waltzed into the backyard for a quick view of Saturn! It *is* viable to set this telescope up in the field for special occasions; I've seen renowned SCT astrophotographer Jason Ware bring his 16 inch LX-200 to the Texas Star Party and assemble it with only a little assistance. But for the most part this is an instrument to be used in an observatory setting. In this environment, the 16 inch is capable of doing amazing work. Some of the best amateur sky photos ever made have been produced by the LX-200 16 inch.

The 16 inch is considered a part of the LX-200 family, but in many ways it is really a completely different telescope from the rest of the group. It uses a totally redesigned and *much* more massive fork and drive base. The computer operating this giant SCT is also of a different design and includes some features and modes not found in the smaller telescopes. Most of the changes in the software and electronics allow the 16 inch to be easily adapted for remote control. Despite these mechanical and electronic changes, the telescope's optics are much like those of the 12 inch size. The

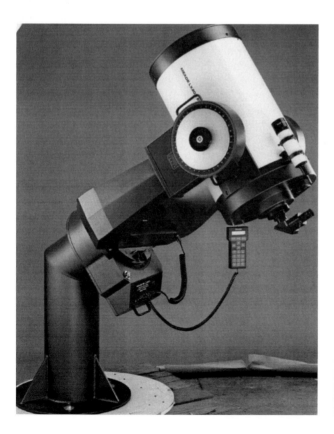

Figure 4.10. The gigantic LX-200 16 inch (photo courtesy of Meade Instruments Corporation).

corrector of the 16 inch is, like the 12 inch lens, made of BK-7 glass. Meade has also limited this scope to f/10 focal ratio optics – no f/6.3 version is available for either of its biggest SCTs.

Celestron's Large CATs

Celestron has offered a variety of fork-mounted large-aperture scopes over the years. Since their introduction in the 1970s, the fork-mounted 11 inch C11 and 14 inch C14 have become familiar and respected amateur instruments. These telescopes were joined in the 1990s by the "almost" 10 inch, the 9.25 inch aperture Celestron Ultima 9.25. But recently Celestron's large aperture philosophy has been changing. During the years when the Ultima goto scope was in development the company began offering its 11 inch and 14 inch OTAs on heavy-duty German equatorial mounts – possibly in an effort to remain innovative by introducing new models of some kind. These configurations, which became very popular with astrophotographers, featured outstandingly good Losmandy G11 equatorial mounts produced by Hollywood General Machining in the US. Not only were these telescopes, the CG11 and CG14, more portable than their fork-mounted predecessors, they were also apparently much more popular with telescope buyers.

The good reception of these SCTs on their G11 mounts prompted the company to begin offering the 9.25 inch SCT on a downsized but similar quality Losmandy-made mounting, the GM8. The growing popularity of all three of these GEM scopes finally lead Celestron to phase out all of its large fork-mounted telescopes. The C14 was the first to go, and was followed in 1999 with the halt of Ultima 11 inch and Ultima 9.25 inch fork-mount production. A further change has been the end of Celestron's agreement with Hollywood General Machining. The 11 inch and 14 inch telescopes are now sold on Celestron-made GEMs, while the 9.25 inch is paired with the imported CG5 mount.

Celestron G 9.25

With the end of the association between Losmandy and Celestron, a new mounting had to be found for what had been called the CG 9.25. Celestron chose to adapt the imported CG5 GEM, which is used on their current G8 SCT, for the 9.25. The 9 inch OTA on this telescope had been an outstanding performer in its previous German and fork-mount incarnations, providing excellent views with its f/10 optics set. It uses a longer "native" focal length mirror (about f/3) than the f/2 primaries found on all other mass-produced SCTs, and some observers believe this helps it deliver superior images.

Unfortunately, the new GEM model of the telescope has some strikes against it. The G5 equatorial mount it's now equipped with isn't rock-solid with the 8 inch OTA onboard, and the considerably heavier 9.25 inch OTA is likely to be even shakier on this mounting. The same lightweight extruded aluminum tripod that was a weak point in the G8 setup is also used here. This large, capable CAT has been saddled with a small and aggravating 30 millimeter finder, too. Does the G 9.25 have *anything* going for it? The price is quite low for this telescope. It's surprising that Celestron can afford to provide even a G5 GEM at this price. As it is, the mount may serve many people well for general visual observing, especially with same fine-tuning of the G5 mount as recommended for the G8 telescope.

CM1100 GEM 11 Inch

The verdict is still out on this telescope's new Celestron-made CI700 mount, since it has only been on the market for a relatively short time, but this new configuration of scope and mount, which Celestron calls the CM1100, does appear to have preserved some of the best aspects of the CG11 while offering some real advantages. The new GEM still utilizes the CG's Celestron-designed PEC drive system, which, like the units on the company's fork-mount telescopes, offers four drive rates. Unlike the fork-mount scopes, however, it also features an expansive set of slewing speeds ranging from $0.3\times$ sidereal for guiding to a $20\times$ rate for short slews. The Losmandy mount was very well respected, but it was hardly perfect. Many users complained about its lack of manual slow-motion controls. In an almost unheard of setup for a GEM, the G11's slow-motions were provided solely by its motors. The CI700 corrects this perceived oddity with the addition of manual slow motions on both axes.

Will the new Celestron mounting prove as popular as the Losmandy? Only time will tell. It is just now getting into the hands of demanding astrophotographers. Early word is that the CM1100 continues to be a stable and highly competent telescope – though some users feel its new CI700 GEM has some rough edges and that the Losmandy was a better mount. Just can't stand the thought of not being able to buy a real CG11? Hollywood General Machining can help. The company can provide a C11 OTA mounted on the original G11 mount. This configuration is produced (or at least integrated) by them, and is not a true Celestron. The Losmandy option does come at a price – it costs considerably more than a stock CM1100 from Celestron.

CM1400 14 Inch GEM

Eleven inches too small? If that's the case, Celestron is now offering its venerable 14 inch OTA on the same CI700 mount now sold with the 11 inch GEM scope. The only difference other than the bigger OTA is the inclusion of an extra counterweight for the declination axis to balance the heavier tube. How well will this big scope work on the CI700? In my judgment, it should perform about as well as the earlier CG14. That is, easily steady enough for visual use, but slightly shaky for demanding imaging tasks. Photography should still be possible as long as it is only attempted on nights when there isn't much wind – even a fairly light breeze may be too much for this pair.

Little CATs

Just as the big jungle cats have cousins in the form of house cats, the big CATs of the telescope world are emulated by little portable telescopes. All SCT manufacturers have offered sub 8 inch SCTs over the years. The most common sizes having been 4 and 5 inches, with an occasional 6 inch also being seen. Originally, the smaller CATs were offered in an attempt to bring the budget-conscious telescope buyer into the SCT fold. Can't afford an 8 inch SCT? How about a nice 5 inch, then? The makers quickly found out, however, that a 4 or 5 inch SCT isn't much cheaper to produce than an

8 inch model and Celestron phased-out the original fork-mount C5 SCT in the 1980s.

With the perceived desire for highly portable telescopes in recent years, Celestron reintroduced the C5. But only a few years after a triumphant comeback, it has again been discontinued. Just before the C5+ was once again taken off the market, it was being sold for *more* than an 8 inch Celestar. Meade has had a similar history with its 4 inch SCT, the 2045. The only genuine small SCT being sold at the time of this writing is the Celestron G5, which is a C5 OTA on an imported German mount. Meade has lost interest in smaller-than-8 SCTs for now, choosing to build its smaller-aperture scopes in the Maksutov–Cassegrain telescope (MCT) design, instead. Fashions and needs change and the little CATs come and go. But there will surely always be a place for small models, whether as true SCTs or as MCTs. As the song goes, "*...but the CAT came back the very next day!*"

Celestron G5

Survival for this endangered species of small CAT has meant giving up its fork-mount. I was somewhat appalled when I saw the first ads for the G5. There was the C5 OTA, whose f/10 optics have always enjoyed a sterling reputation, perched on what looked for all the world like one of those tiny, shaky equatorial mountings you see on the 60 millimeter refractors sold in department stores. What a shame! But when I saw my first G5 in person I was relieved and impressed. The G3 import mount (from China) sold with the telescope is actually much better than it looks in pictures. It provides more than adequate support for the light 5 inch tube, and appears to be very well built. The G3 is equipped with the same light extruded aluminum tripod that causes problems for other CATs on imported GEMS, but the C5 is light and doesn't push this tripod beyond its limits.

The G5 is an amazingly good and economical little telescope. The purchaser gets a high-quality 5 inch OTA on a sturdy mount equipped with a well-designed single-axis battery-powered RA drive. The G5 is supremely portable and stubbornly capable. It can even be made to take good deep sky astrophotos – though this is not an optimum instrument for the budding astrophotographer. The G5 does a lot for a little. It gets a definite and unreserved thumbs-up from me.

SCT-like Telescopes: The Maksutov– Cassegrains

They look a lot like SCTs, but they're a different breed of CAT, the MCTs, the Maksutov–Cassegrain telescopes. Many years ago, in the 1940s, a number of optical designers were searching for a way to do a Schmidt-type telescope without having to produce that extremely difficult-to-make thin corrector plate. Dimitri Maksutov in Russia had an idea. Instead of the complex Schmidt corrector, he'd use a thicker, strongly negatively curved (meniscus) lens instead. After some experimentation by Maksutov and others, this germ of an idea grew into the Maksutov astrocamera, and, a little later, the visually capable Maksutov–Cassegrain. Maksutov's design (see Figure 4.11) as it finally evolved has a number of advantages over the Schmidt–Cassegrain formula.

In the Maksutov, the telescope's focal length is kept reasonably long, about f/15, so all of the elements of the optical system – the primary, the corrector and the secondary – can be left spherical. Spherical optics are, you'll remember, much easier and less expensive to figure accurately than parabolic or aspheric elements. It was also discovered that not only could the secondary, like the primary and corrector, remain spherical, it could become merely an aluminized spot in the center of the corrector rather than a separate secondary mirror.

This all sounds good. Why hasn't the MCT displaced the SCT in the affections of amateur astronomers? There are several reasons. One is focal length. To keep the MCT optical surfaces spherical, the focal ratio must be as large as possible. The first Maksutov–Cassegrains were in the f/20 range. Designers did find that they could bring the f-ratio down to f/15 or thereabouts and still retain good correction. But f/15 is still a long focal length for a visual deep sky observer. An 8 inch telescope at f/15 needs overly long focal length eyepieces to yield the wide-field, low-power views most modern amateurs crave. For a photographer, f/15 is very "slow." Imaging a dim nebula or galaxy at a focal length like this means punishingly long exposures.

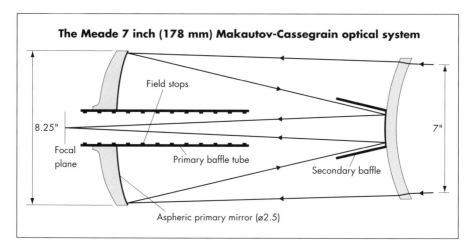

The Meade 7 inch (178 mm) Makautov-Cassegrain optical system

Field stops

8.25"

7"

Focal plane

Primary baffle tube

Secondary baffle

Aspheric primary mirror (ø2.5)

A common MCT problem is tube length. To achieve focal lengths in the well-corrected range, a typical MCT primary has to be somewhat longer in its actual focal length than an SCT main mirror. f/3 seems to be typical (most SCT primaries come in at f/2). This means the tube length of Maks (another nickname for these telescopes) must be longer than that of a same-aperture SCT, causing problems if a fork-mounting is to be used.

A final Mak design deficiency becomes evident if a prospective purchaser is interested in a telescope larger than 7 or 8 inches. In larger sizes the big "simple" Maksutov corrector plate actually becomes considerably more difficult and expensive to make than that of a large Schmidt–Cassegrain corrector plate. The result is that the prices of larger-than-8-inch MCTs are far higher than those of the big SCTs.

With this list of caveats why do we still see Maksutov–Cassegrains? Because the MCT has quite a number of strengths to counter its weaknesses. These telescopes have a reputation for high optical quality. Many of the manufacturers who've made MCTs over the years have sold them as premium telescopes for premium prices. Their customers have expected quality, and these Mak makers have delivered it. It is also true that in reasonable sizes, the MCT optics do seem easier to get right, especially in a mass-production setting. Though the MCT is still an obstructed system – there's a secondary mirror in the light path – these telescopes come very close to duplicating refractor-style performance on the planets.

Figure 4.11. The Maksutov–Cassegrain is similar to the SCT, but with some important differences (illustration courtesy of Meade Instruments Corporation).

What's operating an MCT like? Details vary from manufacturer to manufacturer, but most mechanical designs have been pretty much the same since this scope caught on in the 1950s. If you met an MCT on a dark observing field, it could easily be mistaken for an SCT. These CATs have the same short, stubby tube with the eyepiece in back. And most are on very SCT-like fork-mounts. The majority of MCTs focus just like SCTs with moving mirror systems. Some can even use standard SCT accessories. For the SCT user, the MCT has a very familiar and friendly feel.

In the last five years the Maksutov–Cassegrain telescope has experienced a tremendous surge in popularity. It still hasn't come anywhere near challenging the Schmidt's dominance, but with modern amateurs more interested in high-quality optics than ever (and more willing to pay for quality than ever), the MCT is definitely enjoying a Golden Age.

Questar 3.5

Since its introduction in the mid-1950s by the Questar Corporation of New Hope, Pennsylvania, the Questar 3.5 inch MCT (see Figure 4.12) has been a much lusted after telescope. It's easy to see why. This is a little thing of *beauty*. "Jewel-like" is a good and often-heard description for this telescope. The entire instrument is executed in gleaming stainless steel and beautiful anodizing. You'll search in vain for plastic here! Appearances aside, the 90 millimeter Maksutov–Cassegrain optics are of superb quality, delivering incredibly good contrast and allowing you to view objects and details you'd think would surely be beyond the reach of a 3.5 inch telescope. Though the design of the 3.5 hasn't changed much in 50 years, the innovations found on this small wonder are still fresh today.

A good example is the finder. The 3.5 Questar doesn't really *have* one, not in the usual sense. When a finder for aiming is needed, the observer continues looking through the main scope's eyepiece. Flipping a control on the rear cell switches the eyepiece view to a unique reflex optical system that allows you to see a wide field without moving your eye from the eyepiece (the objective of this finder is mounted on the rear cell underneath the tube). Need more magnification while viewing normally through the main scope? With other

telescopes the user would change to a different eyepiece or insert an amplifying Barlow lens. Not with the Q3.5. Another rear-cell switch places a built-in Barlow in the light path. Again, the observer hasn't had to look away from the eyepiece. Dew falling? Slide the built-in dew cap forward and into place. This is a beautiful anodized tube that fits over the barrel when not extended. Move this dew cap (on which is an engraved constellation map) off the main tube, and an engraved lunar map is revealed on the scope's main tube.

The Questar 3.5 is beautiful and legendary. It looks classic but still modern and sexy. Is this many amateurs' dream scope? Yes. But consider this little CAT's liabilities as well. If there's one thing that keeps me from recommending this telescope wholeheartedly it's the aperture problem. Despite the beautiful silky mechanics, *this is still just a 3.5 inch telescope*. It is a very much *optimized* 3.5 inch telescope, but it can't violate the laws of physics. It will still be outperformed optically by a humble C5. Also, as noted above, the Questar hasn't changed much in 50 years. While the rest of the telescope industry has moved on to high-tech DC

Figure 4.12. The Questar MCTs are beautiful and jewel-like (photo courtesy of Questar Corporation).

drives, for example, the Questar still pokes along with an AC synchronous motor. But what's the main reason that will cause most amateurs to cross the Q 3.5 off their wish lists in a hurry? Price. This 90 millimeter currently retails for nearly US$3,600. Without a tripod.

Still, it's very beautiful. This is astronomy with style. The little Questar is tremendously portable, too. And it's just about as well made and reliable as a telescope can be. This is not a telescope that will often be out of service for repairs or become obsolete in a few years. It is a telescope you'll pass down to your grandchildren. From this perspective it *almost* seems like a bargain. I've wanted a Questar since the 1960s, but common sense (and my budget) has always gotten in my way. In my fantasies, though, I can see myself basking on a tropical isle, cool drink in hand, waiting for a solar eclipse, Questar 3.5 inch at my side!

Questar 7

Everything about the Questar 3.5 inch is also true about the 7 inch. The tube is an upscaled version of the small MCT's OTA. Until fairly recent times, the 7 inch Questar was available only on a fork-mount that was a bigger version of the 3.5's mounting. Today, the Q7 is also available on a GEM mount (made by a third party). The Q7 has always been a rare bird in the amateur community. "Legendary" is again the word often used in reference to this sizable Mak. How does one perform? Beautifully, under the right conditions. For example, the telescope has to cool down sufficiently. Large Maksutov–Cassegrains can require several hours to reach thermal equilibrium and start performing up to spec, and this is definitely the case with the Q7. For best results, the Questar 7 should also be used on objects appropriate for it, too. Large open clusters and nebulae are not its objects of choice. The large focal ratio of this scope, f/15, means that this 7 inch aperture scope delivers high magnifications, with even a 25 millimeter eyepiece producing over 100×.

As is the case with the 3.5 inch Questar, cost is the main barrier between most amateurs and the Questar 7 of their dreams. Think the 3.5 is expensive? This MCT takes us to a whole other realm! The current price for a standard astro Q7 is *over* US$5,000 for an OTA alone. If you want the Questar-supplied GEM or the time-honored fork, plan on spending as much as *another* $5,000! *That's* why these scopes are so rare. In fact, back in the 1960s, these classic MCTs were often called "doctors' telescopes." This was not just because their gleaming stainless steel made them look like a precision scientific instrument found in an operating theatre, but because you'd need to be a wealthy physician to afford one!

Meade ETX

Want a Questar 3.5? Can't afford one? Meade has a deal for you. In 1996, flashy full-color ads appeared in the astronomy magazines heralding the introduction of the Meade ETX (see Figure 4.13). One look at the picture and it was immediately clear what an ETX was: a Questar clone. Reading the copy, it appeared that this little telescope might be even *better* than its inspiration. The telescope's f/13.8 Maksutov–Cassegrain optics were at least a *little* faster than the Questar's f/14.6 speed. The drive also seemed to be an advance over the ancient Questar model, operating from a few penlight cells rather than house current. It was also apparent from the ad that a few of the interesting features of the Questar had been eliminated. Meade did not try to duplicate Questar's reflex finder system, mounting a small 20 millimeter finder on the OTA instead. The built-in Barlow concept was also not attempted. The mounting, though, appeared to be very much in the Questar style, even down to the three little table-top legs which screwed into the fork-mount's drive base like the Questar's. It wasn't surprising that someone could build a telescope similar to the Questar. What was shocking was the price. Meade was asking $500 for its "Questar!"

   You simply can't duplicate something like the Questar 3.5 for a bargain price. Or can you? How does this little telescope compare to the real thing? Optically, very well. I've compared ETXs and Questars set up side by side, and have found that the images delivered by the OTAs are nearly identical. The Questar *might* have a slight edge, but if there is any real improvement in the images produced by the more expensive scope it is extremely slight. Mechanically, the story is completely different. The Questar is built to heirloom quality. The Meade, although fairly sound mechanically, is simply not in the same league. That Questar style fork mount

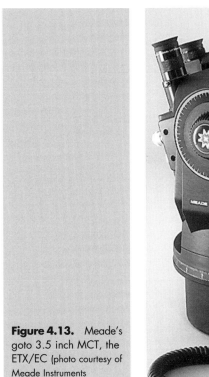

Figure 4.13. Meade's goto 3.5 inch MCT, the ETX/EC (photo courtesy of Meade Instruments Corporation).

and drive base turned out to be made from ABS plastic. Plastic can work well; it is used to advantage in many modern SLR cameras, for example. And it does an OK job as a telescope mount. But the precision and durability of the ETX mounting are not even close to the standard set by Questar.

But I do so love the ETX! For *me*, $500 for a 3.5 inch portable telescope makes much more sense than $3,600 for a very similar 90 millimeter scope with *identical* images. The Questar 3.5 is a joy to use – smooth and wonderfully responsive – but the ETX is fun too. The tiny finder and plastic mounting are irritating at times, but the beautiful images and the low price have convinced me it might be possible to tolerate the little CAT's imperfections with minimal grumbling. The amazing ETX story doesn't end here, however. In January of 1999, Meade announced that there would be a few changes made to the ETX, which would now be called the ETX90/EC.

The new ETX costs about $100 more than the original. What's changed? The RA drive has been replaced by a better unit and a self-contained declination motor has been added. These motors are controlled by an included hand controller much like those found on the big CATs. Four speeds are available from this paddle: $8\times$ sidereal, $32\times$ sidereal, 0.75 degree per second, and 5 degrees per second. At all speeds, the scope's drive works more smoothly than the simple RA unit on the earlier version. Dual motors and a handbox for a hundred bucks more is pretty good value. But what makes the new ETX so amazing is the optional *Autostar*.

The Autostar, which is currently available separately for around $100–$150, looks for all the world like a cross between the LX-200 hand controller and a TV remote control. What does it do for your ETX? It turns it into a *goto scope!* This seemed almost unbelievable at first. For a small additional amount of money you can make your ETX-90/EC into a computerized telescope that will point to any one of 14,487 objects (many of which will be invisible in a 3.5 inch telescope, of course). Most amazingly of all, this little computer system actually *works fairly well*, pointing the ETX to stars, planets and deep sky objects. The Autostar computer also provides an array of functions similar to those you'd expect from any goto telescope: nine speed drive operation, "guided tours" of the sky, extensive information on celestial objects, and a good deal more. Like the LX-200 and Ultima 2000, the Autostar-equipped ETX90/EC can even be operated by an external computer if desired. The earlier RA drive-only ETX is still sold by Meade, but it cannot be adapted for use with the Autostar computer.

In 1999, Meade brought another ETX to market, the ETX125/EC. This is a 5 inch version of the 90, and is, like its little brother, an MCT. This telescope has yet to make the impression on amateurs that the 90 millimeter ETX has. This is because it really is just an upscaled version of the smaller telescope. The plastic that works OK for the 90 renders the 125 shakier and less durable than it could be. Meade is continuing to refine this telescope, though, and the prospect of a fairly large MCT on a goto mounting will be an attractive one once all the bugs are worked out.

Meade 7 Inch MCT

The ETX isn't Meade's first foray into MCT optics. The company has been selling a 7 inch Maksutov–Cassegrain for a number of years. This f/15 OTA can compare very well to the famous Questar 7. "Can" because there have been occasional optical problems with this telescope. When things are right, though, the images are amazingly like those seen in a big, expensive apochromatic refractor. Meade has chosen to offer this telescope in two configurations: on an LX-50 mount or on an LX-200 goto fork system.

Except for modified fork arms to accommodate the 7 inch OTA, these fork/drive bases are exactly the same as those used on the SCTs. The use of these SCT mountings does cause a couple of problems for a 7 inch MCT, however. In order for the OTA with its comparatively heavy corrector to balance correctly, Meade has had to add a lead weight to the rear cell (inside) of the telescope. This significantly extends the already long cool-down time for an MCT of this size. Meade has attached a small fan to the rear cell that helps the cool-down process, but it does not completely eliminate the waiting. The longer native focal length of MCT primary mirrors means that their tubes must be longer than those of comparable size SCTs. The result here is that the corrector end of the tube cannot pass through the fork arms on these Meade mounts. This is mainly a problem during storage, and does not affect the use of the telescope.

The Russians are Coming – the INTES Maks

Actually, when it comes to MCTs, the Russians are *here*. In the last several years, Maksutov telescopes produced in the former Soviet Union have become available to amateurs in the US and Europe. These scopes, most of which have been coming from two sources, Lomo of St Petersburg, and INTES in Moscow, have attracted quite a bit of attention from western consumers because of their low prices and robust build quality.

Thus far, the most popular of these visitors from the east have been those made by INTES. INTES produces a wide variety of telescopes, including a full range of Maksutovs in apertures up to a huge (for an MCT) 16 inches. The INTES scopes are available in several different focal length configurations, too, including f/12, f/10, and the amazingly short f/6 photo version. The biggest seller among these telescopes has been a 6 inch f/12 design, the company calls the INTES MK67.

The first surprise most amateurs experience when they encounter an MK67 is that it's much nicer looking and mechanically sound than they expected. Most of us have seen the mechanically crude cameras and binoculars produced in the Soviet Union and Russia (which can be optically very good despite their appearance). But it's clear at a glance that this little scope is a cut above past Russian export items. The MK67 looks very much as if could have been produced in any western factory. Another surprise for amateurs used to Questars and other traditional Maks is the secondary mirror. The MK67 has a separate secondary mirror in a secondary holder that protrudes through the corrector – just like an SCT. This arrangement of secondary, *Rumak* style, has a couple of advantages. It allows for more flexibility in optical design, and it provides the user with a simple means of collimation. Unlike American made Maks and SCTs, most INTES telescopes abandon the moving mirror-focusing arrangement, using a standard Crayford-type focuser.

If the MK67 is a sign of the kind of telescopes we can expect to see coming out of Russia, western amateurs are in for some real treats. The optics are superb, as good, in the examples I've tried, as what you'll see in any Maksutov.

Celestron C90

Our final Mak–CAT, the Celestron C90, has been around in a number of forms for many years. Celestron has often sold this model as a spotting scope, but has twice offered it specifically for astronomical purposes. During the 1980s it was equipped with a little single-arm fork-mount and called the C90 Astro. This was discontinued after a few years, with the C90 returning to its spotting scope status. The astronomical version was recently revived as a smaller, less expensive mate to the G5. Mounted on the same small German mount as the G5 SCT, this configuration is now known as the G3.

This telescope may be intended as a competitor for the ETX, but it is really a much different scope, and not only because it is sold on a GEM mount. Unlike the ETX, the C90 clearly has a spotting scope heritage. A sign of this is the 0.965 inch visual back – a substandard eyepiece size often seen in the spotters. The G3G3 is furnished with a 1.25 inch American

Standard diagonal/eyepiece, but this is attached via an adapter. Another tip-off is the photo-spotting-scope focusing arrangement. The G3 doesn't use a moving mirror. Instead it uses a simple and camera-friendly arrangement. It is focused by turning its barrel, just like a big camera lens.

The C90/G3 is capable of producing good images, within the limitations of its small 90 millimeter (f/11) aperture. But it's hard to focus the scope to obtain the sharpest images the optics can produce. The turn-the-barrel focusing system simply doesn't do well in high-magnification astronomical use. In addition, while I've seen some nice C90s, I've yet to find one with the exquisite optics of an ETX or Questar 3.5.

Late News: The Celestron Nextars

Nexstar 5

With the introduction of the computer-controlled 90 millimeter and 5 inch ETXs by Meade, many observers of the astronomical marketplace began to wonder about Celestron. How could they let the popular Meade scopes go unchallenged? They couldn't and didn't. Early in 1999, Celestron introduced the Nexstar 5, yet another incarnation of the time-honored C5 optical tube, but this time on a high-tech single-arm fork-mount (see Figure 4.14). It became an immediate hit, and not only because of its excellent 5 inch optics. Unlike Meade, Celestron chose to put its small robo-scope on an all-metal mount, making for a nicer looking and more robust telescope. Most users have found the Nexstar's pointing operation to be somewhat more consistent as well, though Meade continues to improve the software for the ETX's Autostar computer. The Nexstar's attractive design has caught the eye of many amateurs – this little scope looks like it would be right at home on the bridge of the starship *Enterprise*!

Nexstar 8

Just as this book was being completed in the late winter of 2000, Celestron struck again, releasing details of another

Figure 4.14.
Celestron's new goto
C5, the Nexstar 5 (photo
courtesy of Celestron
International).

new goto scope, the Nexstar 8. This will feature the same mount used on the 5 inch Nexstar, but with an 8 inch OTA instead of a C5 tube. I've tested this scope briefly and can say it looks very beautiful, sporting a gray-finished Ultima 2000 style OTA on the futuristic Nexstar mounting. It also seems to work very reliably, and is quite a bit steadier on the Nexstar single-arm fork than I'd feared. I understand that Celestron plans to continue producing the Ultima 2000, aiming the NS 8 at visual observers who want an 8 inch goto telescope for an attractive price and don't plan to do much photography.

How to Buy a CAT: Dealing with Dealers

You know which telescope you want, but how do you get it? I've little doubt that the very best way to buy an SCT or any other telescope is from a local dealer. If your city is blessed with an astronomy shop or a camera store that carries Schmidt–Cassegrains, you're in luck. In addition to being able to see your prospective instrument in person, you'll have peace of mind knowing you'll have someone to help you out locally if something goes wrong or if you merely have a question about setup or use of your new telescope. You'll nearly always pay a bit more when purchasing in-town than you will mail-order, of course. But buying an SCT from a local store is a quick and usually satisfying process.

Are there any reasons *not* to deal with a local merchant? Besides the slightly higher retail prices, local taxes may apply. When you figure in these extra charges, you may find they not only exceed the mail order dealers' shipping charges, but the extra amount adds up to the difference between the model you've chosen and a more elaborate and capable model from a mail-order source. Another all too common complaint about local scope sellers is they may sell Schmidt–Cassegrains, but they often don't know too much *about them*. The large mail-order firms are much more likely to have real amateur astronomers on their staffs than the small chain "nature store." It doesn't do much good to patronize a neighborhood shop if the response when you have a question or problem is always: "I don't know. Guess you'd better call Celestron (or Meade)."

For most of us, buying a new SCT locally is not an option, anyway. If you live outside a major metropolitan area, you're more than likely going to have to order your telescope by mail. This is not necessarily a bad thing. You will, no doubt, save some money, as the big mail scope merchants are highly competitive with one another, keeping prices at rock bottom. Sure, you won't be able to see the SCT of your dreams in person, but one of your local astronomy club's members probably either has the model you're interested in or a very similar type of CAT. Have a question about your new SCT? It's natural to be a bit leery about having to resort to a long-distance call every time you need an answer. But any dealer, local or long distance, is often not the best source of information on telescopes. The book you're reading can answer many questions, especially those of a fairly general nature. But the best source of all is probably your local astronomical society. It's not unusual to find amateurs at an astronomy club with 20 or more years' experience in the care and feeding of CATs. Your town's club is an incredibly good resource on every aspect of amateur astronomy – be sure to take advantage of it.

Shadow of the Past: Buying a Used CAT

Modern Schmidt–Cassegrain telescopes are very reasonably priced, but most of us wouldn't call them exactly cheap, especially when a top-of-the-line model is desired. But there is a way to get a great SCT without breaking the budget: buy used. Buying a previously owned CAT can also allow you to afford a lot more telescope than you could otherwise. For the price of one of today's bare-bones introductory models, you may find that you can have a deluxe scope instead. These older telescopes probably won't feature the computerization and high-tech drives of some of today's more complex models. But they may be fairly feature-laden even by current standards, and may provide you with a good alternative if you want a telescope that's a little more capable of photography than an LX-10 or a basic Celestar.

Another advantage of the used CAT is if you can find a telescope locally, you should be able to check it thoroughly to see if it meets your optical standards. Both SCT manufacturers currently have very good optical quality-assurance programs. But lemons still can and do slip through. If you know just a little about star testing a telescope, and if the seller is willing, you can give a used telescope the "try before you buy" treatment, to ensure that you'll be getting an instrument capable of producing images that will really satisfy you.

What would the perfect used telescope consist of? Ideally, a near-mint condition Celestron or Meade that's only been out of its case a few times over the years. Telescopes like this are, believe it or not, fairly common. What happens is a prospective amateur gets tremendously excited about the hobby, and without doing much – if any – research or thinking, hauls off and buys a big SCT. The telescope gets taken out a couple of times, but interest quickly fades. The owner finds the SCT is too much trouble to set up, that viewing isn't as much fun as he'd thought it would be, or, often the case in these modern times, she just doesn't have the time to devote to a demanding hobby like amateur astronomy.

Almost as good as the closet CAT is the "club circulator." This is an SCT that has been bought and sold a number of times by members of a group. As the needs and interests of astronomy club members change, they sell the CAT to another member. This type of used telescope is probably your safest bet. It will be a known quantity, you'll probably have ample opportunities to try it out, and a number of your fellow amateurs will be intimately familiar with this telescope's operation and history.

What about telescopes from unknown sources? The SCT that shows up in a classified advertisement in your newspaper, for example? There's no reason to avoid these telescopes. Not every amateur joins the local astronomy club – many observers are loners. Just because the person offering the scope hasn't shown up at club meetings, doesn't mean that it won't be a good deal. Some of these telescopes that appear from nowhere turn out to be the biggest bargains of all. If at all possible, though, specify that you be allowed to try the scope in question under the stars before handing over your hard-earned money. Most people who're trying to sell a specialized item like an SCT locally will be only too happy to drag it into the backyard in the interests of closing a deal.

How about buying a used CAT by mail? Caveat emptor! It *is* possible to get a nice scope this way, but the danger of simply buying somebody else's problem increases hugely purchasing a used SCT sight unseen. Most of the amateurs selling their SCTs through classified ads in amateur publications or over the Internet are honest and reliable. But there is some dishonesty out there. And the seller doesn't have to be dishonest for the transaction to turn out to be a disaster. One person's "excellent optics in good condition" is sometimes another's "fair optics, shows a lot of wear." One important safety measure when buying used long distance is to always pay by check. With a check, payment may be stopped if the deal goes sour. If the seller insists on a postal money-order or cash only, be very wary!

After 30 years of mass production, there are now lots of SCTs on the used market. Most are good, but a few are not. Which CATs should be avoided? I advise that those telescopes made during the time of the Comet Halley craze be left alone unless the buyer has the optical knowledge and opportunity to test the SCT thoroughly. During this time (1986–89), both Meade and Celestron struggled to keep up with a demand for Schmidt–Cassegrain telescopes that exceeded even their wildest and most optimistic projections. As a result of this deluge of orders, both manufacturers came close to wearing out their equipment and their workforces and it took them a while to recover.

One common problem to look out for when shopping pre-owned is a lack of enhanced coatings on optical surfaces. Until fairly recently both Meade and Celestron made their enhanced coatings packages for SCT optional. This is one extra-cost option that is more than manufacturer's hype; without these enhanced Starbright (Celestron) or MCOG (Meade) coatings, an SCT's images suffer very real and detectable degradation. The images in a scope without Starbright will appear noticeably dimmer. Enhanced optics are very important in this telescope design, allowing the SCT to make up for light loss in its three optical elements.

Why would anybody order a scope without Starbright? Twenty years ago, SCTs cost about what they do now. But the dollar was worth quite a bit more then than it is today. An SCT was a pretty expensive purchase for the average amateur, and many folks who wanted one were eager to do *anything* they could to keep the price down. The saving involved in buying a non-Starbright scope was very attractive for many.

One of the most important things you can do to help you buy used with confidence is to learn a little bit about the telescope you're thinking of purchasing beyond what the seller can tell you. Information on a particular model's features has often been forgotten by a no-longer-interested amateur or was never known by a casual user. One good way to gain this knowledge is to find advertisements (or even better, reviews) for the telescope in old astronomy magazines. Back issues of *Sky and Telescope*, *Astronomy Now*, and *Astronomy* can be a great resource and are easy to find at club swapmeets. Often your fellow amateurs will be glad to find someone to take part of that mountain of back-numbers off their hands (non-astronomy husbands and wives don't seem to realize the value of a 6 foot high stack of *Sky and Telescope*).

How much should you *pay* for a used Schmidt–Cassegrain telescope? That's a question that I can't answer easily. Prices considered nominal for used telescopes vary from country to country (with used US SCTs being most expensive in Europe). As in many other astronomy-related matters, your local astro-club is your best knowledge base. There is no doubt that the members will be able to give you a good idea of what's a fair price for a particular model of SCT.

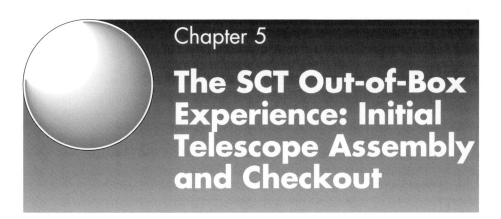

Chapter 5

The SCT Out-of-Box Experience: Initial Telescope Assembly and Checkout

The wonderful day has finally arrived! If you've bought a new SCT, the deliveryman has dropped off several large and impressive boxes. Or, if you're lucky enough to have a local telescope dealer, you've stuffed these boxes into your car. If you've bought used, you've gathered up all the parts and pieces of your "new" scope and brought them home, spreading them out across your living room. However you've obtained your SCT, now's the time to get acquainted. We'll open up those mysterious boxes and assemble your new scope for the first time – indoors. I know you're anxious to get that CAT out under the stars for a look, but it's very important to learn to assemble your telescope during the day when you're rested and have time to figure things out. In the dark, especially if it's cold, everything about your new telescope will be doubly confusing.

The first task is to make sure you have all the components of the new scope. If you've just received a new SCT, you should have two or three boxes. A large box will contain the telescope and its fork-mount. A smaller, tall and skinny box will hold the tripod. The smallest container will hold a wedge if your scope is furnished with one. If any one of these boxes looks damaged, do not let your delivery person leave until you've opened it and examined the contents for problems. Making a claim at this point is much easier than trying to rectify things after the truck driver has gone on his way. In some cases, the boxes may be left on your doorstep if there's no one at home to sign for the shipment. In this event you should call the dealer who sold you the telescope for advice – *not the manufacturer* – if there's any shipping damage. If everything looks to be in one piece, you can send the driver on his way and start unpacking.

Begin with the big box. Be very careful unpacking the telescope. Try not to tear or damage the inner cardboard box – holding the telescope. Why? Because if you haven't ordered a case of some kind, this cardboard box will have to serve as your case until something better comes along. Opening the shipping container will eventually reveal foam cradling your new CAT. Don't let the foam out of your sight. You may need it even if you buy a case for the telescope. Unless you are willing to pay for a very expensive custom container for your SCT, this packing material will be used to line whatever sort of box eventually serves as your case. It is especially important to retain the foam if you intend to purchase Meade's soft case, since the cordura Meade case fits around this original shipping material.

And there it is: the gleaming Celestron Black or Meade Blue tube of your new telescope! If you wish, you can now gently lift the tube and fork assembly out of the box and set it upright on its base. Remove the scope from the container by grasping its fork arms and lifting. Be very careful – it may be heavier than you imagine. You'll probably find that the scope is covered in a protective plastic bag and that various shipping straps or cardboard pieces are attached to the tube and fork. You can remove all this, retaining it for the moment. Observe any cautions or warning notes attached to the tube (often concerning moving a computerized scope in either axis with the locks tightened down). Take a moment to admire your new baby if you wish, but please resist the temptation to play with all those attractive knobs, switches and levers. If you've successfully removed all the packing material from the OTA/fork-mount, you can now gently lift it and return it to its box.

Your next task is to track down and identify any accessories that may be lurking in the main telescope box. Different models and different manufacturers mean items are packed in varying manners. Manu-facturers may ship some accessories in a separate carton, but at least a few of these items are usually included in the main telescope box. A medium-sized container about 4 inches long will contain your star diagonal. You'll also find an eyepiece somewhere. This can be one of several different designs, but it should have a focal length of around 25–26 millimeters. Be sure to locate the all-important hand controller (hand paddle). It will look a little like a television remote control but will be furnished with either a coiled or straight telephone style cord to connect it to your telescope. In a small box or plastic bag you'll find your visual back. This is a ring-shaped adapter with a knurled set-screw on its circumference. Your telescope's 30 or 50 millimeter finder may also be packed in with the main scope. Another common item to find packed in with the telescope is a small tool/hardware kit. This often contains any bolts needed to fasten the telescope to its wedge or tripod and the small wrenches needed for these bolts (often of the "Allen" type). One final and critical item you'll find in the big box is your manual. Because accessories vary with manufacturer and model, check the packing list or other documentation to make sure you've received and located all the items.

Take all the accessories you've retrieved from the container and set them aside in a safe place. The room probably looks as if Christmas morning has arrived with wrapping paper and packing materials everywhere, and the last thing you want is for an eyepiece or a hand paddle to be thrown out with the trash! Once everything's safely set aside, you can tackle the next box.

That skinny box will hold the tripod. Open it carefully so as not to damage the contents, and remove your new tripod. As with the telescope, remove any plastic or other packing materials and set them aside. Examine the box very carefully for any stray items. The tripod leg spreader and, perhaps, fasteners for holding the wedge to the tripod should also be found in this box. Don't return the tripod to its container once you've checked it over, because we're going to be using it shortly. If there are any instructions included with the tripod, set them aside with the rest of your accessories.

Depending on your model and configuration of SCT there may be one container remaining. This will hold the wedge. If you've purchased a computerized goto scope the wedge is not needed for tracking, and will not be included with the scope unless you've specifically ordered it. If you've got a wedge, open this box, remove and examine it. It may be a collection of parts requiring assembly, or it may already be put together depending on the particular model. If the hardware needed for attaching the wedge to the tripod wasn't included in the box with the tripod, you should find it here. Meade telescopes don't normally use bolts to attach the wedge to the tripod, so don't be disturbed if you don't find any. Temporarily retain any packing materials, and place any instructions you find with the telescope's manual.

You should now have all the basic components of your new SCT spread out before you. If you've purchased a used telescope, you should also have all of the items we've mentioned. Actually, you may have *more* accessories if the former owner included the items he or she accumulated over the years in the purchase price. What now? You can tidy up a bit, but, for now, hold onto *all* the packing materials. In the event that something turns out to be drastically wrong with your new telescope, you'll need this material in order to repack the telescope for shipment back to the seller. Chances are that this won't happen to you, but my advice is that you retain boxes and packing for at least a few weeks after the delivery of your scope.

Can we start putting the telescope together now? Not yet. The next job is to sit down and *read the telescope's manual*. I know this is difficult. You've got a brand new CAT in front of you, and it's terribly tempting to jump right in. Resist the urge. While the instructions for assembly that will be given here are a good general guide to putting your telescope together, they can't substitute for the manufacturer's specific instructions. There are many different SCTs, new and used, out there, and there are quite a few differences in how they go together. If you've purchased an LX-200 or an Ultima 2000, reading the manual is not just desirable – it's *mandatory*. So, grab a cold drink, find a comfortable chair, and settle down with your user's manual.

Have you been able to digest the manual? Do you have at least a general idea of how your telescope goes together? If so, let's dig in. Use the instructions included with the CAT if there's any question about what to do or when to do it. Otherwise, the following procedures should enable you to mount your telescope on its tripod and wedge (if your scope has one), and complete a basic mechanical checkout of the telescope, its mount, and its drive and electronics.

Assembling Tripod and Wedge

The first step, naturally, is to put together your tripod–wedge combination. Most of Celestron's wedges are pre-assembled at the factory so you shouldn't have to do anything other than mount the assembly on the head of your tripod. The Meade wedges typically require minor assembly, usually consisting of bolting the tilt plate to the wedge with included hardware and putting together the altitude adjuster. A complete guide to wedge assembly is found in Meade's current telescope instruction manuals.

As for the tripod, most will also require a little work, but this usually amounts to nothing more than installing the leg spreader. The spreader, the metal bracket that holds the legs of the tripod firmly apart, slides over a threaded rod that hangs beneath your tripod. A knob threads onto the rod after the spreader, and turning this "tension knob" tightens the spreader against the tripod legs. Install the spreader and knob, and tighten it against the legs as instructed by the manufacturer. But do not overtighten. The design of the Meade spreader, especially, means that it can break if too much tension is applied. Some Meade designs may require that you attach the wedge to the tripod before the spreader can be tightened into place. Once you've turned the knob enough so as to have the spreader snugged against the legs, make sure the legs are all spread completely apart. If they are, you're done. If they're not, move them until they're as far apart as they'll go, and retighten the spreader.

If your tripod has extendable legs, leave them fully collapsed for now. You may, however, want to check the locking bolts or levers that are loosened to extend the legs. Tighten them so the legs don't extend suddenly when the tripod is moved around, but, like every other part of the scope, don't overdo it. Hand tighten only.

Unless you have a Celestar or an alt-azimuth-mode telescope, the next thing you'll need to do is fasten your wedge to the tripod. Here, the two manufacturers' tripods diverge a bit. Most Celestron wedges attach to their tripods by means of three bolts, which extend through three slots in the wedge (the slots allow the wedge to be moved in azimuth a limited amount during polar alignment). Place the wedge on the tripod, lining up the holes in the wedge with three holes in the tripod and thread in three included bolts. These bolts, in most cases, have Allen-type heads, and can be tightened with a large Allen wrench that should have been included with the telescope. For now, you may tighten all three bolts securely. If we were setting the scope up in the field, the bolts would be left slightly loose to enable you to move the wedge from side to side to align on the celestial pole. SCT users leave their wedges attached to their tripods most of the time.

If you have a Meade telescope, place the wedge onto the threaded rod that extends through the tripod head. Follow the instructions in your telescope manual to properly position this rod. The wedge is then fastened to the tripod by screwing a single large knob onto the protruding end of the threaded rod. The Meade system is very convenient, because it allows you to attach the wedge to the tripod with one simple fastener. Securing the wedge in this fashion gives unlimited motion in azimuth when the knob loosened. This is very handy during polar alignment.

Mounting Telescope on Wedge

This is the moment you've been waiting for; this is where you actually take your beautiful new telescope out of its box and set it up for all the world (or at least your family) to see and admire. If you have an equatorial style setup with wedge, you'll be mounting the scope onto the tilt plate of the wedge. You'll see two holes and one large slot (at the top) in this flat plate that correspond to three holes in the drive base of your telescope. If you are the proud new owner of an LX-200 or Ultima 2000, you'll probably be setting your telescope up in the alt-azimuth mode at this point, which means that you'll mount it directly to the head of your wedgeless tripod.

Let's tackle equatorial telescopes first. To begin, set your SCT upright in its box and check to see that the declination lock is engaged. If you're not sure, try to gently move the scope between the fork arms. If you can move the scope easily, the lock is not engaged. We want the lock "on" to ensure the scope tube will not

Figure 5.1. Thread one bolt into the base a few turns.

flop around while you're lifting it. If the scope seems secure in declination, you can now check the RA lock. *Partially* engage the RA lock. The aim is, as with the declination axis, to make sure the telescope does not move in RA as you're lifting it. But you want to leave the RA lock only partly engaged to prevent damage to the gears if you accidentally bump the drive base on something. When the locks are properly positioned, lay the telescope back down in its container with the control panel side of the drive base *facing down* in the box.

Locate the three bolts that are used to secure your telescope to the wedge. On most recent models these bolts will all have knob heads, but certain earlier telescopes (especially Celestrons) will have allen-head bolts that will require a wrench for tightening. Place two of these bolts in your pocket and take one over to the case. Thread this bolt a few turns into the hole in the drive base that is on the side facing up (see Figure 5.1). Thread the bolt far enough in so it's secure, but leaving it far enough out so that it can slip into the slot at the top of the tilt plate. About three full turns will usually do the job. Before you do anything else, walk over to the tripod wedge, and make doubly sure that all the screws in the wedge, particularly those in the tilt plate, are fully tightened as per the manufacturer's instructions.

When you're ready, go over to the telescope box and grasp the SCT firmly by the fork arms. Carefully lift the telescope out of the box and walk it over to the tripod. The control/power panel of the telescope should be facing away from you. Position yourself at the tripod so you're facing the wedge's tilt plate. Lift the SCT and put it on the wedge by sliding the base onto the tilt plate and the single bolt into the slot on the top, curved part of the tilt plate (see Figure 5.2).

With the CAT now resting on the wedge, you can tighten the bolt to secure it. Leave the bolt just slightly loose so you can move the drive base to line up the holes in the base with the two remaining holes in the tilt place if necessary. To facilitate tightening the bolt, you may loosen the RA lock and turn the fork so that one of the arms is facing up (toward you as you face the tilt plate). Pulling gently toward you on this arm as you tighten the bolt makes things a little easier.

You can now insert the other two bolts through the tilt plate and into the drive base. Meade wedges usually have a set-screw arrangement in the slot at the top of the tilt plate that helps ensure the holes align in the up/down direction. Celestron wedges may require you to slide the scope up and down on the tilt plate a little in order to align the holes. If needed, you can loosen the single bolt holding the scope to the wedge, but please be careful *not to loosen it so much that*

Figure 5.2. Slide the base of the SCT onto the wedge tilt plate.

there's a danger of the scope falling off the wedge! Secure the two remaining bolts firmly, but do not overtighten.

Mounting an alt-azimuth-style telescope onto its tripod is a similar, but usually easier, operation, since you don't have to mount the scope on a tilted surface. Begin, as with the equatorial SCT, by checking the declination and right ascension axis locks. This time, however, make sure both the declination and RA levers are *not* fully locked. You must be very careful not to move either axis with the locks fully engaged, as this can cause very severe damage to the motors and gears within these complex telescopes. The declination lock can be very slightly engaged to prevent unwanted motion, but should never be snugged down tight during this operation.

Lift the scope by the fork arms, just as with the equatorial CAT, and carry it over to its tripod. At the tripod, just lower the telescope (carefully) onto the tripod head. Align the drive base with the head of the tripod as accurately as possible. Some scopes will have a base that is the same diameter as the tripod head, making this an easy task, and some won't. If you're mounting an LX-200, you'll now simply screw the tripod's threaded rod into the drive base in accordance with the manufacturer's instructions. Celestron's Ultima 2000 fits onto its tripod in a similar fashion, but is secured with mounting bolts that are threaded into the drive base from the underside of the tripod. A central alignment pin on the top of the tripod head makes it a simple task to get the light Ultima 2000's drive base holes lined up with the holes in the tripod.

The new CAT is now resting safely and securely on its tripod, and you can begin the basic mechanical and electrical checkout of the telescope. You may, if you wish, take a short break and admire your new companion. Impressive, isn't it? It looks almost like a thoroughbred horse, champing at the bit, ready to race off among the stars! That will come soon enough, but first let's install accessories, and make sure they and the telescope all work exactly as advertised. Let the instruction manual that came with your telescope be your main guide. If there's any question about what to do or how to do it, especially when it comes to the drive, follow the included instructions from your manufacturer. In the case of the computerized LX-200 and Ultima 2000, follow the manual to the letter, performing the initial inside checkout of the telescope (if any) exactly as directed.

Take one more look at the scope for any possible shipping damage or mars and blemishes that may have escaped detection by factory QA inspectors. Loosen the declination lock, and move the tube until it's roughly horizontal. Lock the declination lock securely once you've correctly positioned the tube. This will give us easy access to the front and back of the OTA. The paint job on the tube should be good, with no deep scratches or any other problems. Nothing should be dented anywhere. Fasteners holding together the tripod and other components should be tight and secure and there shouldn't be any obvious cracks or other defects in metal castings. If you haven't done so already, remove the lens cap from your SCT and take a look at the optics.

Optical Inspection

What should your optics look like? Looking through the corrector plate, you'll see the primary mirror at the bottom of the tube. It should be bright and shiny with no blemishes or other defects visible. The corrector should also appear clean and free of scratches, dirt or other contamination. This is a good time to mention what Meade refers to in its manuals as "the flashlight test." If you shine a bright flashlight or other intense light source down the tube, you'll immediately think the optics in your Schmidt–Cassegrain are a *total loss*. The corrector and the primary mirror will look *horrible*, appearing to be a welter of dust, dirt and scratches. Things look especially bad if you hold the flashlight off to the side and shine the light in at an oblique angle. One glance and you'll be ready to return the new telescope to the place where you bought it!

But this is an illusion. Shining a bright light in this fashion will make the optical components, especially mirrors, of any telescope appear to be a mess. What happens is that a substantial amount of light is being scattered across the surface of the mirror. Even the best coatings cannot reflect 100% of the light hitting them. A large part of the remainder is dispersed across the mirror's surface and causes insignificant and nearly microscopic flaws and dust to stand out in stark and *frightening* relief. To correctly evaluate the condition of your mirror and corrector just examine them in room light. A new telescope's optics should, when viewed under normal light, look clean and evenly coated. If they do not, call the seller. If you have a used telescope, it's possible that the corrector needs cleaning. If you think this might be the case, proceed to the "maintenance and troubleshooting" section of this book and follow the directions found there for cleaning optics.

It is not uncommon to find a few specks of dust, paint or other material clinging to the inside of your telescope's corrector plate. Usually these have become dislodged from the inside of the tube during shipment and have landed on your lens' inside face. A few bits of dust and dirt will have *absolutely no effect* on your scope's performance, and you really shouldn't worry about them. If, however, you think there's *excessive* dust and dirt clinging to the inside of the corrector plate, you should call your dealer for advice.

What else can be seen by looking down the corrector end of our new CAT? One thing is the secondary mirror holder in the center of the corrector plate. It should be firmly attached – it shouldn't be possible to move it from side to side in the corrector. If the holder does not have a protective plastic cover, the three screws that allow adjustment of the secondary mirror for collimation of the telescope will be visible. Don't do anything with these screws at this point; just note their location for future reference.

Move around to the rear of the OTA and check out the focusing mechanism. Before trying the focus control, refer to your instruction manual to find out whether the telescope has a "shipping bolt" holding the primary mirror stationary that must be removed. These shipping bolts are generally found on larger CATs, 10 inch telescopes and above, and usually only on Meade telescopes. Their purpose is to prevent the large and heavy primary mirror on these scopes from flopping around during shipping. Attempting to focus the scope without freeing the mirror can cause *severe* damage to the focusing mechanism.

Once the primary mirror's free, exercise the focus knob. The knob on today's SCTs is invariably a silver-colored cylindrical affair mounted on the rear cell not far from the rear port. Older SCTs may have different looking knobs, but they work the same way. Locate this focus control and turn it back and forth a few turns. It should move smoothly without any binding.

If the focus control won't turn at all or binds, there's a problem. One possible reason is that the mirror is, for some reason, at the upper or lower limit of its travel. See if the control will turn in the opposite direction. Turn the knob gently, and, as always, never force anything. If this doesn't produce any results, double check that the shipping bolt has been removed if the telescope has one. If the focuser still won't turn it's time to call the seller for technical support.

If all seems well with the focus mechanism, take a look at the CAT's rear port, the hole in the back of the rear cell where you'll mount eyepieces and other accessories. It should be covered by a plastic cap. This cap usually just snaps off and is not threaded on. Remove it by gently slipping it off and take a peek through the rear port and up into the innards of the SCT. You'll be looking up through the baffle tube, and should see the reflection of the telescope's primary mirror in the secondary mirror. The surface of your secondary should, like that of your main mirror appear clean and free of blemishes, scratches and dust when observed in *normal light.*

Installing Accessories

Leave the cap off because you're going to begin installing accessories now starting with the *visual back.* Go over to the collection of scope components you've set aside and locate this item. One end of the visual back will be equipped with a loose, threaded ring. Place this end against the raised and threaded lip of your SCT's rear port and thread the visual back onto the scope by turning the ring (see Figure 5.3). Take care not to cross thread the back and do not overtighten. With the visual back in place, you'll be able to attach other 1.25 inch accessories to the scope.

You *could* insert an eyepiece directly into the visual back and view "straight through." But that wouldn't be very comfortable. It's easy to see that you'd really have to contort your body into some uncomfortable positions to look through the eyepiece when the scope is pointed anywhere near directly overhead. SCT manufacturers provide us with star diagonals to alleviate this problem. The star diagonal is inserted into the visual back ahead of an eyepiece, allowing for comfortable viewing even when you're pointed near the zenith.

Figure 5.3. Thread the visual back adapter onto the rear port.

Go back to your pile of SCT accessories, locate your star diagonal, and prepare it for use by removing the lens caps (if any) on both ends. Return these to the box or bag the diagonal came from so you don't misplace them. Examining the star diagonal, you'll see that one of the tubes which protrudes from the prism (or mirror) housing has a raised lip and a set-screw. This is where your eyepiece goes. The other tube, at a 90 degree angle to this, is bare and usually chrome plated. This is the end that goes into the visual back. Insert this end into the back and tighten the set so the diagonal is held securely. If you can't seem to insert the diagonal, back-off on the visual back's set-screw a bit. Always avoid loosening these set-screws completely, though. They have a tendency to drop into the grass or thick carpet and disappear forever if you're not cautious. You'll note that you can loosen the set-screw slightly and rotate the diagonal to various viewing angles. This will allow you to position the eyepiece into the most comfortable position possible.

You can now add an eyepiece to the diagonal. We won't be viewing anything, but the presence of an eyepiece will help balance the scope, allowing us to check the declination and right ascension motions. Remove the lens caps from the eyepiece and put them somewhere where they won't be lost. Insert the eyepiece into the open end of the star diagonal and secure it with the set-screw.

Really starting to look like a telescope now! The next step is to install the finder scope. Without a finder in place, even with a star diagonal and eyepiece attached, many SCTs are off balance and tend to swing around wildly when the RA and declination locks are released. Locate the finder and the ring part of its bracket (if this has not been installed on the OTA at the factory). The little finder scope slides into the rings in the bracket and is held in place by three set-screws on each of the two rings. Slide the finder into the rings now. Don't worry about adjusting the set-screws – we'll have to loosen them to align the finder outside. For now, just tighten the two sets of screws so that the finder is held roughly in the center of each ring

If your finder bracket is not attached to the OTA, refer to your instructions for guidance. Usually, a separate finder bracket is attached to the OTA with a quick release setup. This is a bracket on your OTA that has two set-screws for holding the finder/mount in place. Slide the finder's ring mount into this bracket and tighten the set-screws. If your bracket is not of the quick release variety, you'll attach it to the telescope with a couple of screws that thread into corresponding holes on the OTA's rear cell.

Initial Mounting Checks

Check the RA and declination movements of the telescope next. If you have a computer-driven model, please review your instructions to see if it's permissible to turn the slow-motion controls with the power off (in most cases it is). Exercise the right ascension axis first. Grasp one fork arm, and holding firmly so the scope doesn't swing free, release the RA lock. Turn the lock to the right (usually, see your manual) until it is completely disengaged. You should be able to move the scope easily and freely around the drive base. This RA motion should be smooth and free of *any* binding. Rotate the scope 360 degrees until you're back at your starting position and reengage the lock, but only part way. Move it about halfway back to the left until the telescope no longer swings wildly. With this amount of tension on the RA axis, you can try out the slow-motion control, the knob located very close to the RA lock.

You'll recall from our earlier discussion that this control can *only* be used when the RA lock is at least partially *disengaged.* Trying to turn the slow-motion control with the lock fully on will cause severe gear damage, rendering the slow-motion control inoperative and requiring a trip back to the factory for the drive base. So don't do it. Lock the scope only tightly enough so that it doesn't move by itself before operating the slomo control.

Turning the knob gently should cause the telescope to move in right ascension. If the control seems too hard to turn, don't force it; loosen the lock a little and try again. Be careful not to completely release the RA lock suddenly. With the telescope mounted on its wedge, it may swing free and hit you. Even with a finder, star diagonal and eyepiece in place, many modern SCTs are still corrector-end heavy. This is to help the SCT maintain balance with heavy items like cameras attached to the rear port. When you've exercised the RA slow-motion control enough to see that it's working properly, return the RA lock to its completely locked position.

Locate the declination lock and disengage it to about the halfway point. While doing this, keep your hand on the OTA in case you accidentally unlock the dec lock all the way and the scope swings free. This is more of a problem with the Celestron CATs. With them, unlocking the dec axis allows the telescope to move completely freely. Meade telescopes, even with the declination lock all the way off, retain a little friction on this axis, and require some force to move the OTA in declination. How do you move the scope on this axis? If there's a handle on the back of the rear cell, it's easy; just grasp it and move the OTA up and down. If there's no handle, the most convenient way to move the scope is by grasping the tube toward the corrector and rear cell end. Place one hand near the corrector and one on the rear cell and push gently in the direction you want to go. Please keep your hands away from the corrector plate itself. It seems natural to grab the lip of the tube, but if you don't have a dew shield in place this will result in a corrector lens covered with fingerprints by the end of the evening. Move the OTA up and down in declination and pass the corrector end through the forks. The scope should move smoothly in declination with no binding or sticking. When you're done, return the OTA to its original and roughly level orientation and engage the declination lock all the way.

Unlike the RA slow-motion control, which must not be operated unless the RA lock is at least partially disengaged, the declination slow motion won't work *unless* the lock is on. Before trying the declination slow motion, make sure the tangent arm that drives it is roughly centered in the fork arm so that the scope has some travel room. Identify the fork arm with the declination slow-motion control. On most scopes, the dec slow-motion control is at the base of the fork arm that includes the declination lock. Usually, SCTs have two chromium knobs on either side of this fork arm. Look at the inside of this arm. You'll see a threaded shaft that runs through it. The slow-motion knobs are attached to either end. A threaded block rides along this shaft and is attached to the tangent arm that runs up to the declination bearing at the top of the shaft. You turn a knob, the rod turns, the block moves along it, and the tangent arm is moved forward or back causing the scope to move in declination. If you keep turning a slow-motion knob in one direction, the block will eventually move to the end of the threaded shaft and into the fork arm. When this happens, you'll have to reset the declination tangent arm (see Figure 5.4).

Look at the declination slow-motion mechanism. Are the block and the arm roughly centered in the fork arm? If so, all is well. If not, center them. Begin by disengaging the declination lock. Be careful of the OTA swinging free when you do this. Either hold it in place

Figure 5.4. The declination slow-motion assembly.

with one hand or gently lower it between the fork arms. With the lock off, turn one of the slow-motion knobs until the block and arm begin to move in the proper direction. You'll notice that with the lock off, the OTA doesn't move at all when you turn the control. When the block and arm are centered, return the OTA to its horizontal position and lock the declination lock securely.

With plenty of tangent arm travel room, you can test the telescope's declination slow-motion movement. Start turning the slow-motion control in one direction or the other. Note that the tube begins to slowly and smoothly raise or lower. Continue turning until the OTA reaches the end of its slow-motion travel in one direction. When the tangent arm approaches the furthest extent of its travel forward or back within the arm, the control may become slightly more difficult to turn. This is normal. But, in general, movement of the OTA and the slow-motion control should be smooth and easy with no sticking or jamming. When the arm comes to the end of travel, stop, and begin turning in the other direction. Movement this way should also be smooth and free. When you reach the end of travel in this direction stop and turn the control the opposite way until the tangent arm is centered again, and the scope's tube is back in its horizontal position.

Drive Checks

Initial checkout of the new telescope includes one more very important item: testing the clock drive and hand controller for proper operation. For this check, you'll want to follow the procedures in your manual closely to avoid damage to the telescope's electronics. Start by locating the hand controller among the accessories. Bring it over to the scope and identify the proper receptacle to plug it into on the telescope's control panel. Be certain you have the correct port. Some telescopes have several similar-style receptacles on their control panel. Plugging your hand unit into the wrong socket could potentially damage the telescope's electronics. The correct port should be labeled "Handbox," "Hand Controller," "HBX" or something similar. Refer to your manual. Having located the right place for the controller plug, check that the telescope's main power switch is in the "off" position. *Never* plug anything into or unplug anything from an SCT with the power turned on. "Hot-plugging," as it's called, can cause fatal damage to telescope electronics systems. Now, plug the hand control into the appropriate port.

Before going any further, the telescope will need a source of electrical power. Many modern SCTs are powered by internal batteries. If your new CAT uses this power source, ascertain the type and number of batteries required and the location of the battery compartment. What sort of batteries should be used with an SCT? The only recommended type is alkaline. Rechargeable batteries usually don't maintain a sufficient voltage for telescope systems. Plain old lead batteries are deficient in the power department, and won't last long if used in today's power hungry CATs. The majority of current CATs feature battery compartments in their drive bases that are easily accessed via plastic covers. Some older SCTs required you to remove the control panel by unscrewing screws to replace batteries, however. Meade's ETX series of scopes have battery compartments located on the underside of the drive base. When inserting AA cells, be sure to observe correct polarity (polarity means which direction the plus and minus ends of the cells face).

If your DC powered telescope does not work off internal batteries, you'll have to furnish a separate power source. Most amateurs use a small lead-acid battery, one designed for use in lawn and garden tractors, snowmobiles, or motorcycles. You'll also have to trot down to your local electrical/electronics supplier and obtain an adapter that consists of a pair of alligator clips on one end and a female cigarette lighter receptacle on the other. The alligator (or "spring") clips are attached to the terminal posts of the battery. Red clip to positive and black to negative. With these attached, you can plug the telescope's 12 volt cord into the adapter's receptacle.

A battery of this type is inexpensive and can work well. Just take extreme care to always get the red clip on the positive post and the black clip on the negative post. A help here is to paint the base of the battery's positive post red and the base of the negative post black. Don't paint the part of the terminal post where the clip attaches, though, as paint is an insulator and will prevent electricity from flowing to your drive electronics. If you choose to use one of these storage batteries, try to obtain a sealed type. This will eliminate the danger of an acid spill if the battery is accidentally turned over in the field.

Some telescopes will be equipped with AC power supplies as well as 12 volt DC cords. If your scope has one of these, you can plug it into the wall for our drive testing and worry about batteries later on. You'll still need a battery, of course, if you plan to operate your telescope away from home at remote dark sites.

If you've purchased a used older telescope with an AC synchro drive motor, you won't need a battery. You may have a drive corrector or an inverter for your scope, but leave them aside for now. All we want to do at this point is test the basic operation of the telescope drive. Just plug your telescope into a household AC receptacle. All of the AC drive CATs I've seen lack an on/off switch. If you want the drive to run, you plug the scope into the wall socket.

With batteries installed or an AC line cord ready to be plugged in, we now come to the moment of truth. Will the drive run? Will the electronics be OK? Turn on your telescope's power switch or plug it into an AC receptacle. On most battery-powered telescopes you should see an indicator light. Lights on the hand controller may flash as well, indicating that the telescope is performing a self-test. The AC drive telescopes usually don't have a light of any kind, so your only indication that all is well will be the gentle hum of your motor running. If, on the application of power, nothing seems to happen, or worse, a grinding noise or smoke results, immediately turn the power off. If nothing happened, check your connections carefully. Batteries inserted in the proper orientation? External 12 volt cord secure? Alligator clips firmly attached to battery posts in the proper polarity? If these checks don't cure the problem, if again nothing happens when you apply power, or if you get noise or smoke, stop immediately. It's time to call the manufacturer or seller of your telescope.

Assuming that the initial power on sequence went well, proceed to a cursory indoor check of the drive and hand controller for proper operation. As a first test, press the east and then the west buttons. This will cause the drive to change speed. Even at 2 times the drive rate, which is what pushing these buttons will usually deliver, movement of the scope will be very slow. Your best indication that the controller east/west buttons are working properly will be a slight change in pitch of the motor's sound when either button is pressed. If your scope is equipped with a declination motor, you can try the N/S buttons as well as the E/W buttons. It will be easy to see that the declination motor runs, as you'll be able to detect the declination slow-motion shaft turning slowly.

If you've received an add-on declination motor but haven't installed it yet, turn off your telescope's power and install it per the manufacturer's instructions. This is usually a fairly simple operation, with most external dec motors simply slipping over one of the declination slow-motion knobs. You may have to add a small motor bracket to the fork arm and change one of the stock declination slow-motion knobs for another one that will be included with the dec motor.

You can hear the RA motor turning, but is anything happening? Watching the movement of a telescope as it's driven at the sidereal rate is like watching the hands of a clock – literally. But if you wait for a few minutes, you should be able to detect that the scope has moved. If, even after you've waited a considerable length of time, no movement on the RA axis is detectable, check the right ascension lock. Remember: *it must be engaged in order for the RA drive to move the telescope.* If you live in the southern hemisphere, this might be a good time to set up your scope for proper movement. Consult the instructions for advice on how to select southern hemisphere operation for the drive. With "south" selected, the drive should move the scope in the opposite direction from that of the "north" setting.

If you're the happy owner of a battery-powered telescope, your drive probably offers a number of slewing speeds and other capabilities you'll want to try. How you activate the telescope's higher speed motions is dependent upon the model. With some, it will require a combination of button pushes on the hand controller. Others will have you push a speed selector button or reposition a switch. However it works, check your scope for proper movement at the higher speeds. Some telescopes, particularly Celestrons, feature an assortment of drive rates different from, but similar to, the sidereal rate. You won't be able to detect a change in the movement speed of the scope, but you should see an indicator light up on the control panel showing that you're at "King," "lunar," or "solar" drive rate.

At this time, you can also test any other electrical indicators or accessories for proper operation.

If the control panel includes a power meter, it should be indicating a value representing the power consumption of the drive. If the hand controller has a map light, it should light when you turn it on. Focus motors and other accessories, like illuminated reticle eyepieces that are powered from the telescope control panel, should be working properly. Follow the instructions provided with these devices for assembly, and please remember to turn the main telescope power off before plugging them into their ports on the panel!

GEM Telescopes

All of the preceding applies to fork-mounted telescopes. What if you've purchased a German mount version like a G8 or a CG11? There will be a few differences from fork-mounted telescopes, naturally. Assemble your GEM mount as per manufacturer's instructions. Remove the OTA from its shipping container, mount it on the GEM and check all of the telescope's and mount's workings just as we did with the fork-mount.

Are there any potential problems that might be encountered during the assembly of a German equatorial-mounted SCT? Putting a GEM SCT together involves sliding the telescope tube onto dovetail bar on the mounting. You slide the telescope on from one end and secure it with set-screws. This is not a difficult task with an 8 inch telescope. But with an 11 inch or a 14 inch this can be a frightening experience the first time you do it. You have to lift that big tube and maneuver it into place. Work slowly until you get accustomed to mounting the OTA. You might want to have a family member or friend help you those first few times.

Any other "gotcha's" involved with the GEMs? Counterweights. A GEM's counterweight slides over the end of the declination shaft and balances out the OTA, which rides on the other end. Always mount the counterweight onto the shaft before putting your OTA on the GEM. Otherwise, especially if you've neglected to tighten down the mount's locks, the telescope may swing wildly in right ascension, stopping when your beautiful tube smashes into the tripod or mount. When the time comes to disassemble the telescope after a night of observing, always remove the OTA from the mount *before* the counterweight for the same reason.

You will have to play with the position of the counterweight in order to balance the telescope correctly in RA. You slide the weight back and forth along the dec shaft with locks/clutches disengaged until you find a position where the scope achieves balance. A GEM also needs to be balanced in declination to work properly. To do this, you move the scope tube itself forward and back in the mounting's dovetail bracket or saddle until balance is good on this axis. Take extreme care that the scope does not slide off the mount while you are attempting to achieve declination balance!

Often, German mounts are sold without drives. The motors are options. You'll be required to install the mount's drive motors the first time you set up the telescope. This is not difficult if you follow the instructions that came with your CAT (assuming the instructions are at all clear – those included with small imported mountings often leave a lot to be desired). The main thing to watch out for is that the RA and declination slow motions must move smoothly before you install the motors. If the axes of your mount don't move easily, the motors will stall and possibly burn out. Imported mounts may require slight adjustments of the gears and additional lubrication before they're ready to go. If in doubt, contact the telescope's maker for instructions.

Telescope Disassembly

This concludes initial testing of your new scope. Turn the power off, rest for a moment, and maybe step back and take a last long and admiring look at your new baby. If everything has gone well, disassemble the telescope by removing it from its wedge and tripod and returning it to its container.

Begin by unplugging the hand controller and removing any accessories like declination motors and focus motors that you may have attached. If the finder has a quick release bracket that must be detached for the scope to fit in its case, remove this as well. Replace the cap on your scope's corrector. With all the subsidiary items removed, position the scope in RA so that it'll go into its case correctly. The fork should be parallel to the control panel. When this is done, tighten the RA lock part way (unless you've got a goto model). Unlock declination and move the OTA to its storage position with the corrector end of the tube between the fork arms and pointing toward the base. On non-goto scopes, you can lock the declination lock all the way. To remove the telescope from the wedge, unscrew the two lower bolts from the tilt plate. *Leave the top bolt, the first one you inserted, in place.* Loosen this bolt *just slightly* so you can slide the telescope off the wedge. Grasp the scope by the fork arms and slide it off the tilt plate, sliding the single remaining bolt out of its slot. When the scope is free, carry it over to its case or box and gently lower it into place with the control panel side of the base down and the finder side of the OTA up. If you are dismantling a computer-goto scope, undo the knob or bolts holding it to the tripod, lift it straight off the head, and carefully return it to its box.

You can remove the wedge from the tripod if you wish by reversing the assembly process. Normally, though, there's not much reason to do this. A tripod with a wedge on top doesn't take up much more space than a tripod without one.

That's all there is to it. Now just wait for the Sun to set and it's time to set the CAT outside for its first night under the stars! I know you can hardly wait for some spectacular views through your beautiful new telescope, and it seems as if it'll take forever for darkness to arrive. But this waiting time can be put to good use by discussing some additional accessories. Despite the seemingly wide array of accessories and gadgets that came with your new SCT, there are some additional items you'll need, items that can be critical to getting full use out of your SCT. There are also quite a few products that, although not absolutely necessary, can make observing with your CAT much more enjoyable and productive.

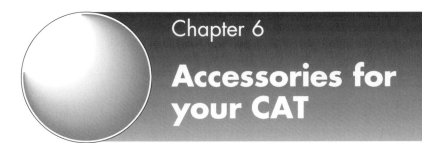

Chapter 6

Accessories for your CAT

You'd think that after the substantial investment in money you've made in a new Schmidt–Cassegrain you'd have everything required to begin a rewarding observing program. Unfortunately, this is not the case. Because of the need to keep the prices of CATs as low as possible, telescope manufacturers have really cut back on included accessories in recent times. Today, the standard SCT doesn't come with much more than a single eyepiece (often of modest quality) and a 1.25 inch star diagonal. You'll quickly find there's some additional buying to do. And there are a fair number of extras that are vital for getting the most out of your telescope.

Eyepieces

At the top of the list of required extras are eyepieces (also referred to as *oculars*). Sure, the telescope came with *one*. But that's not nearly enough. A single eyepiece doesn't allow for much flexibility in observing. A decent selection of eyepieces delivering a wide range of magnifications makes observing much more fun and productive. If you're like many of us, a general interest observer, you may find yourself viewing large open star clusters one night, the Moon the next evening, and a planet or two on another observing run. The star cluster would call for low magnifications, the Moon for medium/high powers, and the planets usually demand high-magnification oculars.

Where does the beginner start? How many eyepieces do you need? What kind? You already have one eyepiece, the one that was included with your scope. Most of today's Schmidt–Cassegrains are shipped with a 25–26 millimeter ocular. Recently, due to competitive pressures, it's become common to find a high-quality Plössl eyepiece in the box with a new telescope. With an f/10 8 inch SCT this eyepiece delivers 80× (magnification = telescope focal length/eyepiece focal length: 2,000/25 = 80). A 10 inch f/10 CAT gives you 100× with the same eyepiece. This is a nice medium power, and most Plössls will perform very well. Even a modest Kellner (SMA) will do fairly well in this 25 millimeter focal length.

It's also a good idea to have at least one eyepiece in the long focal length/low power range. Good 40 millimeter focal length oculars can be had for very reasonable prices and will provide 50× with an 8 inch telescope, a good "finding" or "scanning" magnification.

Looking at small deep sky objects or the planets calls for a relatively high power ocular. A 12 millimeter eyepiece yielding 160× makes a nice medium/high-power eyepiece for the beginner equipped with an 8 inch SCT. If you plan on doing serious planetary observing, though, you'd be wise to choose a slightly shorter focal length eyepiece as your high-magnification ocular. A 10 millimeter gives 200× with an 8 inch telescope, and is just about perfect for delivering planetary details.

A reasonably versatile and inexpensive beginner's eyepiece assortment, then, might consist of 40 millimeter, 25 millimeter and 12 millimeter eyepieces. For an f/6.3 focal length telescope, a 25 millimeter, a 12 millimeter and a 6 millimeter will provide a comparable range of magnifications.

But what *kind* of eyepieces should the CAT owner buy? A look at an astronomy vendor's catalog reveals an almost bewildering array of designs. Luckily though, the many eyepieces inhabiting manufacturers' show-rooms are rather easily broken down into three simple categories: *good, better and best*. What quality of eyepieces are best for use with an SCT? Buying *the very best* ocular that can be afforded is never a mistake. This may be a good idea even if it means getting along with only a few really good eyepieces for a while. There's not much sense in having a case full of eyepieces if they're all of poor quality. Wise amateurs concluded a long time ago that the quality of the eyepiece is *just as important* as the quality of the rest of a telescope's optics. It doesn't pay to invest in a premium telescope only to ruin the image it produces with an eyepiece that would be more at home on a 60 millimeter department store trash-scope. Remember, too, that eyepieces can be used on any telescope that's purchased in the future. A good-quality eyepiece may seem expensive, but it can be used for a lifetime of observing pleasure.

Sticking to high quality does not mean a good eyepiece collection has to empty the bank account, either. Today's amateur is lucky in that there are many good quality eyepieces available for reasonable prices. Before surveying what's out there in the world of eyepieces, though, it's necessary to explain a few terms that every eyepiece buyer will encounter. There are five very important specifications listed for every eyepiece: its *focal length*, its *eye relief*, its *true field* its *apparent field* and its *barrel diameter*.

Focal length indicates how much magnification an eyepiece will deliver. Longer focal length eyepieces deliver lower magnifications than shorter focal length eyepieces do. Use the formula (telescope focal length/ eyepiece focal length = magnification) to calculate a given ocular's power in a particular telescope. Which focal lengths of eyepiece are commonly available? Common focal lengths run from about 40 millimeters down to around 6 millimeters. There are both longer and shorter focal length eyepieces, but these tend to be special and more expensive items.

Eye relief is simply a measure of how close your eye has to be to the eyepiece's eye lens (in its simplest form an eyepiece is composed of an eye lens on one end of the barrel and a field lens on the other) in order for the entire field of view to be visible. An eyepiece with very short eye relief can be very uncomfortable to use. You have to jam your eye right up against the eye lens to take in the eyepiece's whole vista. If you're an eyeglass wearer and need to wear your glasses while observing – if you suffer from astigmatism, for example – you may not be able to see much at all. Your glasses will keep your eye far enough away from the eye lens that you only see a very small portion of the center of the field.

Two more important and related eyepiece character-istics are *true field of view* (also called actual field) and *apparent field of view*. True field is easy to explain. It is the expanse of sky visible through the eyepiece expressed in angular degrees. If, for example, a given eyepiece just fits the entire disk of the Moon into the field of view, it has a true field of 0.5 degrees or 30 minutes of arc. Why? Because the Moon itself is half a degree across. If only half the Moon fits into the eyepiece, then its true field is 0.25 degrees or 15 minutes of arc. The true field of an ocular will vary depending on the focal length of the telescope it's used with. Longer focal length scopes produce higher powers and smaller true fields with a given eyepieces.

What about apparent field of view (AFOV)? This is a term you hear bandied about by amateurs quite frequently. Beginners often find it a difficult concept to understand, but it is really quite simple. Apparent field is the diameter of the circular field of an eyepiece expressed in degrees. If you're told that an eyepiece has an apparent field of view of 50 degrees, this does *not* mean that it'll show you a swath of sky 50 degrees across. A 50 degree 25 millimeter eyepiece shows you something less than 1 degree of sky. The situation here is analogous to the size of a television screen. Comparing an eyepiece with a large AFOV and one with a small AFOV is like comparing a 12 inch screen television set showing an image of the Grand Canyon to a 27 inch screen-size TV showing the same scene. The large TV shows a much more expansive, easy to view version of the landscape. Using a large AFOV eyepiece is like viewing the universe through a large spaceship porthole rather than a little peephole. Modern eyepieces usually have AFOVs from around 40 degrees (inexpensive Kellners and Orthoscopics) to 80 degrees or more (super-premium wide-field eyepieces).

One final specification (and a very important one) for eyepieces is barrel diameter. The vast majority of the oculars you'll encounter have barrels of the American Standard size. This diameter, 1.25 inches, was practically the only size used by serious amateurs for many years. It is now common, though, to see eyepieces in barrel diameters of 2 inches as well. These big oculars are popular for a couple of reasons. One important justification for these impressive looking eyepieces, is that, due to the restrictive size of its barrel, a 1.25 inch eyepiece can only deliver a true field of view *so* wide. At a certain combination of short telescope focal length and long eyepiece focal length, there will be vignetting. You won't see the entire field the eyepiece/scope combination is capable of delivering – it will be cut off by the small diameter barrel.

Another reason for the proliferation of 2 inch designs is the physically large size and heavy weight of ultra-wide-field eyepieces. Many of these premium lenses have special dual barrels that allow them to be used in either 1.25 inch diameter or 2 inch diameter focusers or star diagonals. They are usually much more secure when used in 2 inch units, however, because of their weight (around *2 pounds* for some examples).

You'll occasionally see small-diameter *Japanese standard* eyepieces. These 0.965 inch barrel diameter lenses are rarely seen these days except on very cheap telescopes (the ubiquitous 60 millimeter refractor). Some Japanese manufacturers, like Takahashi, do still make extremely high-quality 0.965 inch barrel eyepieces, though.

"Good" Eyepieces

These oculars are the workhorses of the astronomy world. They aren't the most sophisticated designs, but when used so as to emphasize their strengths, they can do a very good job. Just about every amateur has a few of these in his or her eyepiece case, and most observers continue to find uses for them even after they begin to accumulate premium oculars.

Kellners

The Kellner builds on an earlier and now mostly obsolete design of eyepiece, the Ramsden. In fact, you'll still occasionally hear Kellners referred to as achromatic Ramsdens. The Kellner, which has been around since 1842, is composed of three lens elements. Apparent field of view is a little restrictive in these eyepieces, usually topping out at about 45 degrees. The Kellner does for the most part deliver bright, sharp edges especially in longer focal lengths. Eye relief in these longer focal lengths is quite generous.

Many new amateurs are very happy with their long and moderate focal length Kellners, especially if they've previously had the experience of using the terrible little eyepieces that are included with department store refractors. So they buy another Kellner, usually a higher power, shorter focal length model, and are very disappointed. There's a sad fact of life about Kellners:

they perform well *only* in longer focal lengths. Shorter focal length Kellners tend to exhibit poor edge of field sharpness and often exhibit increased spurious color. In my opinion, the practical limit for focal length of Kellners is about 12 millimeters. Shorter than that and these eyepieces really fall apart.

When is a Kellner not a Kellner? When today's telescope companies choose to call them by other names to give them a panache of quality not associated with the Kellner name. You'll see eyepieces described as *modified achromats*, *super modified achromats*, and other fancy-sounding titles. But they are all humble Kellners. If you see the word "Achromat" in the ocular's design name, think "Kellner," because that's what you'll be getting.

Erfles

The Erfle, with its more sophisticated five-lens-element design, is definitely a step up from the Kellner. Not too long ago, Erfles were considered premium eyepieces and were one of the first really good oculars the new amateur gravitated to. Today, the Erfle is still respected, though it has been far surpassed by more modern ocular designs. Actually, the Erfle was the first truly modern eyepiece, being formulated for use by the world's militaries in World War I. What was needed was a low-power, wide-field eyepiece that was sharper and had better color correction than the simple and poor-performing Huygenian, which was the most popular eyepiece at the time for low power applications. The Erfle has occasionally been produced in shorter focal lengths, but its real domain is in the 25 millimeter plus range.

The Erfle promises better performance than the two and three element oculars and it delivers well on this promise. Particularly striking for the observer used to the restrictive apparent fields of Kellners is its expansive apparent field of view. Many Erfles approach AFOVs of 70 degrees or more. Color correction is also generally good. So why don't we hear more about the good old Erfle these days? There is essentially one reason why the Erfle design has been superceded by the modern wide-angle and ultra-wide-angle eyepieces – edge of field sharpness. Erfles do offer a wide field, but the stars close to the edge of this field tend to look more like blobs or seagulls than stars. The Erfle also often exhibits problems with internal reflections. Looking at a bright star or planet generally produces one or more ghost images in one of these eyepieces. Not that an Erfle is a bad eyepiece by any means. An Erfle is a good compromise for the observer who wants a wide-field, low-power ocular but doesn't want to spend a lot of money.

"Better" – The Semi-Premium Eyepieces

Orthoscopic

When I was a young amateur, the one eyepiece my peers and I wanted more than any other was the Orthoscopic. This was as good as you could get in the 1950s and 1960s, and we would have trashed all our humble Ramsdens and Kellners in a moment for a single Ortho. Alas, the price was much too high for us. Ironically, in these latter days Orthoscopics are the most reasonably priced of the premium eyepieces, usually selling for appreciably less than the currently more popular Plössls.

What is an Orthoscopic like? It's a four-element design dating back to the mid-nineteenth century. It is also known by its original name, Abbe, after its designer. It picked up the Orthoscopic name, however, in recognition of its amazingly distortion-free field. And except for a couple of brands (Zeiss still calls its eyepieces Abbe-Orthoscopics), this is the name these eyepieces bear today. Orthoscopics are usually found at the opposite end of the focal length scale from Erfles. There are longer focal length Orthoscopics produced, but generally their "beat" is the 25 millimeter and *down* range.

What kind of a job does an Orthoscopic do? A very respectable one. Their apparent fields are nothing to write home about, usually being in the 50 degree range. But this one weakness is easily offset by the design's strengths. The difference between a short focal length Orthoscopic and a Kellner is like night and day. The stars in the Orthoscopic are little pinpoints even close to the edge of the field. Color correction is good, eye relief is fair, and most Orthoscopics hold onto all their good qualities even when used with fairly short focal length telescopes. The Orthoscopic makes a wonderful high power eyepiece for the amateur who wants quality on a budget.

Plössl

The Plössl is now, without a doubt, the most popular eyepiece for amateurs. It, like the Orthoscopic, dates from the mid-nineteenth century. Also like the Orthoscopics, it is a four-element design. Why has the Plössl, which is also sometimes called the *symmetrical*, become so much more popular than the Orthoscopic? The major reason is that modern eyepiece makers have chosen to experiment with this design, producing modified Plössls that perform exceedingly well, even with today's fast short focal length reflecting telescopes. The Orthoscopic, on the other hand, has remained much the same as it has ever been.

But even a stock Plössl design has some real advantages. It is relatively free of internal reflections, even in longer focal lengths. Its color correction is quite good, and its edge of field sharpness is good across the entire range of focal lengths from long to short. It is, as a result, just as likely you'll see 40 millimeter Plössls as 5 millimeter Plössls. The apparent field of the Plössl tends to be slightly better than that of the Orthos, but not by much, usually maxing out at around 50–55 degrees. In one way the Plössl can actually be slightly worse than the Orthoscopic – eye relief. If you wear glasses, beware of some shorter-length Plössls; in short focal lengths the eye relief of some brands of these eyepieces can be smaller than that of a comparable Orthoscopic ocular.

The "Best" – Modern Wide-Angle Designs

These are ultimate tools for the modern amateur: those big, heavy exotic-looking wide-field eyepieces. The trend toward large and expensive eyepieces was started as long ago as the 1980s by Al Nagler's company, Tele

Vue, with its Nagler and wide-field eyepieces. This movement toward very well corrected eyepieces with large apparent fields was soon picked up on by Meade, who debuted its Ultra Wide and Super Wide oculars not long after. In the late 1990s, it has become well accepted that modern observers will pay as much for a dream eyepiece as a photographer will for a good zoom lens, and the market has expanded further. Japanese optical giant, Pentax, offers its own line of super-premium eyepieces, the XLs, and another famous Japanese optical manufacturer, Vixen, is also on the scene with its Lanthanum Superwides.

Is one of these expensive eyepieces for you? Only you can answer that. Take a look at our survey of these oculars below, ask fellow amateurs what they think of the big eyepieces, and try a few for yourself at star parties. One warning: after a single viewing session with a spacewalk eyepiece, you'll find it very difficult to go back to the relatively tiny apparent field of a Plössl or an Orthoscopic, and your wallet will surely suffer!

Tele Vue Naglers

These are the amazing eyepieces (see Figure 6.1) that got the spacewalk eyepiece craze started. This series of oculars is currently available in focal lengths of 4.8, 7, 9, 12, 16, 17 , 20, 22, and 31 millimeters. These eyepieces are also identified by their types. There are plain Naglers (the 4.8 and 7), Type 2s (the 20, 16 and 12), Type 4s (12, 17, and 22) and the new Type 5 (31 millimeter). The type represents a design differ-ence. The Type 5 is the most recent design; there were apparently no Nagler Type 3s. As this series of eyepieces has evolved, the offered focal lengths have changed from time to time, with an 11 millimeter and a 13 millimeter disappearing some years ago. While the Type 2s are still widely available, I expect them to be phased out in favor of the more sophisticated Type 4s.

How good are these Naglers? I can unreservedly say the Naglers are *great* eyepieces. With the Type 4s coming on the scene to correct for the single major failing of some of the Naglers, deficient eye relief, there's frankly nothing bad I can think of to say about

them. In fact, I'll go so far as to state that my own 12 millimeter Type 2 is without doubt *the best eyepiece I've ever used.* Period. The new replacement for the Type 2 12 millimeter, the Type 4 12 millimeter is reputed to be even better! Sharp, and I mean *tack sharp,* amazingly wide vistas are what these eyepieces offer. The 17 millimeter, 22 millimeter, and 31 millimeter Naglers require a 2 inch star diagonal, but the rest can be used in 1.25 inch or 2 inch units.

Figure 6.1. Tele Vue's Naglers (photo courtesy of Tele Vue Optics, Suffern, NY).

Meade Ultra Wides

Meade currently offers four ultra-wide-angle eyepieces (see Figure 6.2). Their focal lengths are 4.7, 6.7, 8.8, and 14 millimeters. So, the focal length range is a little different from that of the Naglers. How do the Meades compare to the Tele Vues in other ways? In quality, I'd say that the Ultra Wides are almost as good as the Naglers. They are *very* fine oculars and provide the same expansive viewing experience found in the Tele Vue eyepieces. If edge of field sharpness and eye relief is not quite as good as in the Naglers, this may be because Meade has not chosen to update its designs noticeably. The Tele Vues have evolved, but the Ultra Wides have

remained much as they were when they were introduced in the 1980s. They are *extremely* nice oculars, however, and can certainly make even the most critical observers happy. It is also nice that their focal lengths are at least slightly different from those of the Tele Vues.

Tele Vue Panoptics

The Panoptics (see Figure 6.3) are the descendants of Tele Vue's original line of "almost Naglers," the wide fields. The Panoptics feature apparent fields slightly narrower than those of the Naglers, about 68 degrees versus the Naglers' 82 degree range. Comparable focal length Panoptics are also slightly less expensive than their Nagler cousins. One attractive feature of the Panoptic series is that its lower focal length range is wider than that of the Naglers. In addition to 15, 19, and 22 millimeter Panoptics, Tele Vue also sells 27 and 35 millimeter versions. This is nice for low-power eyepiece buyers (including many MCT owners with long focal length telescopes) who want Nagler-like wide fields.

Figure 6.3. The Tele Vue Panoptics (photo courtesy of Tele Vue Optics, Suffern, NY).

Meade Super Wide Angles

Tele Vue sells "almost Naglers" and Meade, not to be outdone, markets "almost Ultra Wides." These are the Meade Super Wide Angle oculars. They bear the same relationship to the Ultra Wides as the Panoptics do to the Naglers. Like the Panoptics, Super Wides boast generous, but not huge, 68 degree fields. Also like their Tele Vue competitors, the Supers are available in a focal length range that includes some low power models. The current Super line-up consists of 13.8, 18, 24.5, 32, and 40 millimeter oculars. Criticisms? These are not quite as nice as the Panoptics. The design for the Super Wides, like that of the Ultra Wides, has remained frozen in time.

Tele Vue Radians

These are the new kids on the eyepiece block (see Figure 6.4), having been introduced in 1999. Although these eyepieces (available in focal lengths of 18, 14, 12, 10, 8, 6, 5, 4, and 3 millimeters) are certainly priced within the premium range, their 57 degree fields (in math, one radian is equal to about 57 degrees) wouldn't seem to place them with the rest of these high-priced eyepieces. What makes them so special that Tele Vue can ask this much money for them? Simply put, many observers think these are the best planetary eyepieces *ever made*. They are sharp. Incredibly sharp, outdoing some of the legendary planetary eyepieces like the Zeiss Abbe-Orthoscopics handily. All focal lengths feature

Figure 6.4. Tele Vue's new Radian eyepieces (photo courtesy of Tele Vue Optics, Suffern, NY).

20 millimeters of eye relief, which makes this eyepiece easy to use. A person who must wear glasses while observing can, because of this large amount of eye relief, still see the whole field of view while wearing them. With short eye relief eyepieces, the eye of a person who wears glasses is too far back from the lens for the whole field to be seen.

Pentax XLs

For quite a while, Meade and Tele Vue had the wide-angle market to themselves. But this is changing. Other manufacturers are catching onto the fact that big apparent field eyepieces are "hot" in the current amateur market. Long-time and well-respected Japanese camera maker Pentax has made quite a splash in the amateur market with its SMC XL series. These five-seven element oculars are made in focal lengths of 5.2, 7, 10.5, 14, 21, 28 and 40 millimeters. Apparent fields of 65 degrees for most of the series (55 degrees for the 40) make them more comparable to the Panoptics and Super Wides than to the Naglers, but in price they are very similar to wider field Naglers and Ultra Wides. What's the attraction then? They have quickly gained a reputation for being of very high optical quality. The availability of short focal lengths has also caught the attention of planetary observers (who don't need 85 degrees, anyway).

Vixen Lanthanum Superwides

In addition to being a famous telescope maker, Japan's Vixen has long been a producer of high-quality eyepieces. Its original line of Lanthanum oculars was acclaimed for delivering 20 millimeters of eye relief in very short focal lengths (even in the 2.5 millimeter focal length model) and was quite popular despite fairly narrow AFOVs. Vixen was therefore moved to introduce another series of Lanthanums, which kept the 20 millimeters of eye relief but improved upon the restrictive fields of the originals (about 45–50 degrees). Vixen also dramatically improved sharpness and reduced aberrations, both problems with the first Lanthanums.

The Lanthanum Superwides are a series of 65 degree apparent field eyepieces that are comparable to the Tele Vue Panoptics in price and performance. All of these eyepieces can be used in 1.25 inch star diagonals, since they are equipped with 1.25 inch/2 inch barrels. Another plus for the Lanthanums is that they are available in much shorter focal lengths than the Panoptics or Meade Supers, making them nice tools for Lunar and planetary workers. At this time, Vixen is manufacturing these eyepieces in focal lengths of 3.5, 5, 8, 13, 17, and 22 millimeters.

Eyepiece Cases

As you begin accumulating eyepieces, you'll soon discover you need a place to put them. Some kind of box or case to keep your eyepieces safe from bumps, dust and dew is essential. A number of manufacturers sell cases especially made for this purpose, but, actually, any type of container filled with protective foam padding will do. Particularly good are some of the hard cases sold in camera stores to hold photographic equipment. These hard-sided, brief-case-sized containers are usually furnished with cubed foam. This is foam padding that has been cut into small cubes. You remove these cubes in order to form a custom recess for your eyepiece or other accessories. Some of these photo cases can be a bit expensive, but if your eyepiece collection consists of high priced, super-premium eyepieces, you'd be well advised to provide them with the best protection you can find.

Other Optical Accessories

Barlow Lenses

What is a Barlow lens? Something every single amateur should have in the eyepiece case. What does it do? In its simplest form, a Barlow is a single-element negative lens that increases the magnification of eyepieces. A Barlow is an *eyepiece amplifier and multiplier*. It works like this: you install a Barlow in your star diagonal in the same place the eyepiece usually goes. You then insert an eyepiece into the *Barlow*. When you look into the eyepiece you find its power has been doubled (usually). If that 25 millimeter eyepiece gave you 80× with your scope, it will give you 160× when "Barlowed." The 25 millimeter now gives the same image scale and true field as a 12.5 millimeter eyepiece. Another way of looking at a Barlow's effect is that it multiplies the focal length of your optical system. If you've got an f/10 telescope, you'll have an f/20 after putting a 2× Barlow into the star diagonal. It's not just a focal length or eyepiece amplifier either, it's an eyepiece *multiplier*. If you own a 25 millimeter and a 15 millimeter eyepiece, the addition of a Barlow lens effectively doubles your number of oculars, giving you focal lengths of 25 millimeters, 15 millimeters, 12.5 millimeters and 7.5 millimeters.

What should you look for in a good Barlow? It should be well corrected for spurious color, incorporating two lens elements into its design rather than just a simple negative lens. That is, it should be an achromatic Barlow. It should also have well blackened lens edges and the inside of the barrel should be baffled or at least painted a really flat shade of black to help prevent the reflections that can devil Barlows. It should be sturdily constructed and have a good set-screw to hold your valuable eyepieces securely. What amplification factor should you choose? Barlows are available in strengths of from 1.8× all the way up to 5×. The CAT owner would do well to stick with 1.8× or 2×. The higher-power Barlows like the Tele Vue's 5× Powermate will usually deliver too much magnification with an SCT or MCT, even when used with longer focal length eyepieces.

Focal Reducers and Reducer–Correctors

Barlow lenses have been in amateurs' accessory kits since the nineteenth century, but the opposite of a Barlow, a focal reducer, is an item that has really only gained widespread popularity among SCT users in the last few years. A focal reducer is a special lens element which, when placed near the focus of your SCT, reduces the focal length of a telescope by a set amount. Put one of these negative Barlows ahead of an eyepiece and the telescope is no longer an f10 – it's an f5 or an f6.3 (the two most popular focal ratios produced by these units). For a photographer, being able to take pictures at a speed of f/6.3 instead of the slow f/10 is wonderful. Deep sky exposure times are reduced in length by nearly two-thirds; this alone is a huge boon for the deep sky imager. Another benefit of the reducer for astrophotographers is the expansion of the scope's field. An f/10 SCT can't be called a wide-field instrument by any stretch of the imagination. The true photographic field of the average SCT is a bit more than 1 degree. OK for many subjects, but far too narrow for big open clusters or huge nebulae. A reducer can *almost* make an SCT a wide-field telescope, nearly doubling this photographic field of view.

"But I'm not a photographer," you say. The focal reducers are very handy tools for the purely visual observer, too. Like Barlows, they can give you a new set of eyepieces for free! The 26 millimeter eyepiece that produced 80× when used with your scope at its normal f/10 focal length gives a pleasantly low 40× with an f/5 reducer in place. A couple of premium eyepieces, a good Barlow, and a reducer–corrector may be all most amateurs will ever need!

Focal reducers have been available for many years. The ones sold for SCTs are usually simple, single-element positive lenses that covert an f/10 SCT to f/5.

How well do these devices work? Fairly well for photographers. They may suffer various optical problems like vignetting (cutting off the field of view) or deficient edge of field sharpness. But in many cases, the edge of the field where these problems occur will be outside the edge of a 35 millimeter film frame and will not trouble astrophotographers at all. When one of these devices is used for visual observing the results can be a little disappointing, though, with seagull-shaped stars at the field edge being the norm.

Early in the 1990s, Celestron and Meade began marketing special focal reducers they called *reducer–correctors* (r/c). This is a special type of focal reducer (see Figure 6.5) that consists of a two-element lens contained in a specially threaded barrel. One end screws onto the rear port of the SCT and the other end duplicates the rear port's threads. The telescope's visual back or other accessories are screwed onto the back end of the reducer–corrector. In addition to reducing the telescope's focal length to f/6.3, these devices *also* apply some optical correction. One of the major faults of the Schmidt–Cassegrain's optical design is that its field is *curved*. Stars near the edge of the SCT eyepiece field don't look as sharp as those at the center, no matter how good the eyepiece is. The r/c almost completely eliminates this problem, flattening the field and making images very sharp across the entire visual and photographic fields of view.

Both Meade and Celestron reducer–correctors perform incredibly well. I rarely remove mine from the rear cell of my Ultima 8. The field flattening introduced by these inexpensive little devices is really startling. All

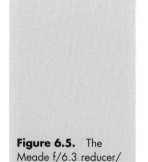

Figure 6.5. The Meade f/6.3 reducer/corrector (photo courtesy of Meade Instruments Corporation)

in all, the visual images on a scope equipped with an r/c are better than they are without it! Any problems with these units? For photographers, they, like the plain focal reducers, may produce some vignetting at the edge of the frame. This is especially bad when the r/c is used on an SCT with an aperture larger than 8 inches. In my own (8 inch) setup, I rarely notice this problem, and when I do, the effect is minor, consisting of a slight darkening of the picture at the negative's corners. Visually, there may be considerable vignetting when using 2 inch barrel diameter eyepieces longer that about 30 millimeters in focal length. This is not a fatal flaw, however, as the short focal length delivered by the r/c makes it usually unnecessary to use eyepieces longer than 30 millimeters.

In my opinion, the first optical accessory a new CAT owner buys after eyepieces and Barlows should be a reducer–corrector. Which one? There are only two versions available, one from Meade and one from Celestron. These two reducer–correctors appear identical and are said to be made in the same Far Eastern factory, so I'd buy whichever is cheaper. The r/c is one accessory where manufacturers' hype turns out to be completely true. This is really an indispensable accessory for every CAT user.

Two Inch Star Diagonals

If you want to use 2 inch eyepieces in your SCT, you'll have to upgrade the stock 1.25 inch diagonal that came with the scope to a 2 inch unit. 2 inch diagonals are available from a large number of companies, and come in two styles. SCT-style star diagonals are made expressly for CATs. They include a built-in visual back – a threaded ring that allows the entire diagonal to screw onto the telescope's rear port in place of the regular 1.25 inch visual back. The other type is what's usually referred to as a *refractor diagonal*. These 2 inchers are enlarged copies of 1.25 inch diagonals. They have a plain drawtube that slides into a 2 inch focuser. Since there's no attachment ring on these, a 2 inch adapter (also known as a 2 inch visual back) is required. This is a tube that screws onto the telescope's rear port. It is of the right size to accept 2 inch eyepieces and accessories and is equipped with set-screws to hold them firmly in place.

Which type of diagonal is best? The SCT style is convenient and compact and keeps the diagonal close to the SCT's back, preventing focusing problems. The combination of a 2 inch adapter and a refractor style diagonal may place an eyepieces so far back that some may not quite reach focus, especially if an f/6.3 reducer–corrector is in use. I still prefer the refractor diagonals, though. To change the viewing angle of an SCT diagonal, you have to first loosen the threaded visual back ring. This is annoying. All you have to do to move the position of a refractor diagonal is loosen one or two set-screws. A much larger selection of 2 inch diagonals is available in refractor format, too.

Dealing with Dew: Shields and Heaters

Maybe you're lucky. Maybe you live in a desert or mountain area, somewhere where the humidity and the dewpoint temperature are always low. But for the rest of us, dew is a major enemy. One of the first accessories a new CAT owner needs is some kind of dew protection unit. Why is dew a special problem for the SCT? Take that new Schmidt–Cassegrain out into the humid night and observe for a little while and you'll *see* why. At first everything is OK. But before too long, you'll notice that the bright stars have developed little haloes. Before too much more time has passed it'll become obvious that images are not as bright as they were a little while ago. Is there something wrong with the new scope? You bet! Shining your red flashlight onto the corrector plate reveals a thick coating of dew. An SCT invites dew formation because of the large corrector plate. Couple this with the fact that you use your scope, usually, out in a wide-open space and that the corrector is always pointed up, and you have a recipe for dew. Dew forms when an object radiates enough of its heat into space to grow colder than the current dewpoint. When this happens, dew "falls."

Dew Shields

A dew shield (or cap) is the simplest and least expensive solution to the dew problem. It is nothing more than a metal or plastic extension to the tube that fits over the front of your telescope. Your finder scope probably already has one. Notice how the objective on this little refractor is recessed from the end of its tube. This forms a dew shield for the small telescope. How does a dew shield work? It helps the corrector plate retain a little heat by *shielding* it from wide-angle exposure to the *heat-sucking* sky. Even if dew is not a huge problem, a dew shield might be a good investment. Eventually you'll transport your scope to an observing site where dew *is* a problem. Nothing is more disgusting than having traveled many miles to set your scope up under a beautifully clear and dark sky only to be shut down by dew formation. A dew shield also serves a second purpose. If you must deal with nearby street lights and other annoying sources of ambient light in your observing location, the dew shield helps keep this contrast-robbing stray light out of your optical system.

An SCT dew shield is nothing more than a simple metal or plastic tube of the right size to slip over the end of the telescope tube. These devices are sold by a number of accessory manufacturers and are also available in the form of flat sheets of plastic that are formed into tubes and fastened onto the end of your scope with Velcro when you're ready to use them. These are convenient because they can be stored flat. A dew shield for an 8 inch CAT currently goes for about US$50.

Dew Guns

You've got a dew cap on the end of your scope. But still it comes. That thick layer of image-destroying dew. You always shut down long before you're tired of observing. The solution? Go to stage two in the anti-dew crusade. If you check the telescope accessory catalogs, you'll find a little device called a *dew gun* or *dew zapper* being sold. This looks just like a little hair dryer, only it plugs into a 12 volt DC source rather than the household mains. And it really is just like a little blow dryer. The idea is that when dew begins to form on your corrector, you direct a little warm air from the zapper onto it until it warms sufficiently for the dew to evaporate. These things really do work. They don't apply a whole lot of heat, but not much is needed to deal with dew. They do have a few minuses for the telescope user. When dew gets really heavy, you may find yourself zapping every few minutes. This gets old in a real hurry. Your observing is constantly interrupted and you soon give up. Also, these intermittent blasts of hot air can cause some image problems. Not only will you be looking thorough heat waves, the blast of heat may cause the corrector to deform slightly. It won't take long for it to resume its original shape, but as soon as it does, dew will begin to form again!

Can't find a zapper for sale at a local telescope merchant or by mail? Check boating and automotive supply houses; these devices are often sold as window defrosters. Can't locate one of these either? In a pinch, a plain old hair/blow dryer can work if there's a source of 120 volts AC available. Just don't set your wife's or sister's dryer to high and pump 2000 watts of heat onto the corrector plate! The SCT will take quite a while to cool down after this! Using the lowest setting and holding the blow dryer a good distance from the corrector plate are recommended!

Dew Heaters

If your area has a dew problem, you'll soon tire of using a hair dryer or a zapper gun. The next stage in our fight to eliminate damp correctors is the use of a special heating element that warms the telescope's corrector plate all the time, keeping it absolutely free of moisture. These devices consist of a narrow cloth band long enough to wrap around the corrector end of a particular size of SCT. Embedded within this cloth strip is a heating element that may consist of resistors or of resistive heat rope. It is fastened in place by Velcro closures and is plugged into either a 12 volt battery or a household AC source.

These heaters work incredibly well and they are really the final answer to keeping dew off from the corrector. Even without a dew cap, a properly installed and operated dew heater can keep your corrector plate bone-dry all night long. There is really no downside to using these devices, either. Some people worry about the constant heat applied to the corrector having a bad, deforming effect on this optical element. In practice, though, it just doesn't seem to happen. All the heaters I've seen apply a very gentle heat – not enough, in my opinion, to cause a problem.

Finder Telescopes and Supplemental Finders

Sure, that new SCT already *has* a finder. But unless the telescope is a top-of-the-line model, it is likely too small. A 30 millimeter finder, which is what many budget and middle of the road CATS are equipped with today, really isn't good enough. A too-small finder won't show enough guide stars, stars to help in star hopping to dim deep sky objects. This makes deep sky object location very difficult. In badly light-polluted surroundings a 30 millimeter finder may make it practically impossible to find anything at all beyond the Moon, bright planets and the brightest stars. In the typical heavily light-polluted backyard, the amateur is faced with stretches of sky that are nearly devoid of visible stars. The 30 millimeter finder just doesn't pull enough stars out of this sodium-streetlight-orange void to serve as guideposts to deep sky targets. A 50 millimeter finder offers a considerable improvement over a 30, revealing *many* more stars, even under quite poor skies.

What are the characteristics of good larger finders? Good optics, of course. There's not much sense in increasing the size of the objective if the views are so poor that there aren't many more stars than can be seen in the telescope's stock 30 millimeter. Luckily, all the 50 and 60 millimeter finders I've seen for sale in recent times have had nice optics. Meade, in particular, makes a very high quality but inexpensive 50 millimeter finder scope. Ring mounts for finders are usually sold separately. Be sure to get a well-made mount designed expressly for use on an SCT.

The Telrad

When is a finder not a finder? When it's a *Telrad*. Some years ago, the late U.S. amateur astronomer Steve Kufeld got tired of finder scopes. He felt they were just too hard to use. The typical finder offers a relatively narrow field of view, and the images are usually upside down. Steve hit upon a better idea: a TELescope Reticle Aiming Device, a Telrad! The Telrad is a finder, but it is not a telescope. Through clever use of a red LED-illuminated reticle and a beam-splitter window, the

Telrad seems to project a bullseye on the night sky. The three concentric circles that form this bullseye represent angular distances in the sky of 4 degrees, 2 degrees, and half a degree.

Looking into the Telrad from the rear, you see these circles floating among the stars. Aiming your telescope is now simple and intuitive. *No upside down images to figure out.* You just put the bullseye in the right place among the stars, and, almost magically, you'll find your target in the eyepiece. The Telrad mounts on a rectangular plastic base that is affixed to your scope tube by means of included double-sided tape. The unit is aligned with the main scope by means of three adjustment screws on the rear, which control the position of the reticle on the window.

If the Telrad is so great, why hasn't the standard finder telescope gone the way of the dinosaur? Despite its clever design, the Telrad is not perfect. One of the biggest problems of this device is that it can be almost useless for observers living in heavily light-polluted areas, because the Telrad is not a telescope. It doesn't gather any light at all. Looking through its beam splitter window, all you see is the same field of stars you normally see with your naked eye. It's fine to say that you just place the reticle in the proper place in relation to background stars, but the urban observer may not *see* enough naked eye stars to properly position the Telrad.

Solar Filters

If practiced safely, solar observing and imaging can be one of the most rewarding fields for an amateur to study. Old Sol, especially at the height of a solar (sunspot) cycle, is endlessly fascinating and can show the practiced observer something new every day. How does the CAT user get started on the Sun? By first ignoring the advice found in many older books on observational astronomy. These books will advise you that *projection* is the safest way of observing the Sun's disk. Projection is easy; place an eyepiece in the telescope, hold a white card or other screen a suitable distance behind it, and focus the image projected on this screen by the eyepiece. Why not do this? The problem is that the closed-tubed SCT or MCT heats up *very* quickly when the unfiltered Sun is allowed into the

OTA. Soon the temperatures brought on by the Sun's fury can rise high enough to cause severe damage to the telescope's internal workings. The secondary mirror holder can possibly warp or even melt, the baffle tube can be damaged, and lubricants can evaporate and condense on the mirror and corrector. *Don't use a CAT for solar projection!*

Instead, use a filter. I do not mean one of the inherently dangerous eyepiece solar filters that used to be included with the small imported telescopes! These were designed to be placed near the telescope's focal point over the field lens of an eyepiece, and could heat up quickly and shatter, exposing the observer's eye to the instantly blinding rays of our star. No, what I mean by "filter" is a well made *full aperture solar filter that fits over the corrector end of the telescope*. Today's safe solar filters are made of either glass or Mylar (plastic) with a thin layer of aluminum deposited on the surface. The filter material is contained in a holder that fits snugly over the corrector plate and attenuates the Sun to the point where it is safe to view the disk with unfiltered normal eyepieces. A filter of this kind also eliminates problems with internal heating of the scope.

What should a prospective SCT solar observer look for in a full-aperture solar filter? Optical quality is very important. As mentioned above, solar filters are made with either glass or Mylar substrate. It seems natural to expect that glass filters will be better optically. This, surprisingly, does not seem to be the case. In my experience, *Mylar* filters are capable of producing much sharper images. Despite their wrinkled appearance (tightly stretching the thin Mylar in its holder to eliminate wrinkles actually seems to degrade images), they deliver fine performance.

The main problem with Mylar filters is that they do not produce a realistically colored image of the Sun. Glass filters deliver a yellow or orange Sol, but most Mylar filters deliver a bluish or greenish image. In practice this may not be a problem. As long as appropriate details can be seen, who cares if the Sun is blue? If this is annoying, an appropriately colored eyepiece filter (used *in conjunction* with the solar filter, naturally) can give the Sun a more normal hue.

Please make sure to purchase the right *type* of solar filter, too. Some makers sell both visual and photographic filters. The photographic-grade units allow more light into the telescope in order to make for shorter exposures on film. These must *not* be used visually. The only solar filter you should ever trust your eyes to is one which is rated for visual use and which has been made by one of the long-established companies serving this market. For safety's sake, make very sure that the filter holder fits snugly and securely over the corrector end of your scope. It *must* be tight enough so as not to be inadvertently knocked off by you or by the wind while you have your eye at the eyepiece. Cap or, better yet, completely remove the finder from your SCT to prevent someone from inadvertently looking through it while the scope is pointed at the Sun.

One of these filters can, if used responsibly, provide a lifetime of safe and interesting study of our star. But what if you really become caught up in solar observing and want to go beyond mere Sunspot watching? There are several manufacturers selling hydrogen alpha (Hα) filters. These devices allow you to view the Sun in the red light of hydrogen. Looking at the Sun in Hα reveals a wealth of features invisible in white light. The Sunspots are still there, but you'll also see solar flares blazing away on the disk and beautiful prominences erupting from the Sun's limb. A wealth of other details is also available for the Hα observer and imager.

How do these filters differ from normal solar filters? The most obvious way is that an Hα filter is usually composed of two filters. The first is the *energy rejection filter* that fits over the end of your scope just like your white light filter does. It is rarely a full-aperture filter, though. Usually the filter part of the unit is only 2 or 3 inches across; the rest of the surface is opaque. The filter part is inevitably an off-axis affair. It is off to the side of the holder and out of the way of the secondary mirror. This small filter area is the norm for a very good reason. Hydrogen alpha filters usually only work well with telescopes with *very large* focal ratios. "Stopping down" the CAT to 3 inches almost *triples* its focal ratio. The actual focal length of the scope remains the same, but reducing the aperture makes the effective focal ratio go way up to f/25–f/30, just right for using one of these filter systems.

The second filter in an Hα system is the actual hydrogen alpha filter, a narrow-band device that permits only the light of hydrogen to pass. It is installed ahead of an eyepiece and may need to be used in conjunction with a Barlow lens (to increase the focal ratio of the scope even further). Some Hα filters require the filter be kept at a certain and constant temperature.

This is done with a small heating element. Because of this, Hα filters will usually need an external power source (often 120 volts AC) to operate. The filter usually has a heater control that can be used to vary the temperature of the element. This allows the observer to "tune" the filter. Changing the temperature will enhance certain details on the Sun while suppressing others. Changing the filter's temperature alters its band pass, the range of light wavelengths it transmits.

This sounds exciting! Many new amateurs have seen dramatic pictures of solar prominences and would love to get a look at the wonders of the Hα Sun. Unfortunately, as alluring as the beauties of the Sun in hydrogen light may be, they come at a high price. Even though amateur-oriented Hα filters are now fairly common items, they are still *very* expensive. The cheapest generally go for about US$1,000. The better filters, the ones that will really allow a good look at those dramatic prominences and flares, may cost *several times* this amount.

It's best to wait a while before thinking about obtaining an Hα system. Get a good white light filter, study the Sun in normal light, and do some reading. If you still remain obsessed by the daystar, I'd be the last person to discourage you from delving further into the Sun's mysteries with a hydrogen alpha filter.

Light Pollution Filters

It isn't long before the new deep sky observer, the CAT owner who's most interested in looking at the universe's distant wonders – galaxies, nebulae and star clusters – realizes that he or she has a problem. The marvels that look so beautiful from the astronomy club's dark sky observing site are very washed out and dim when observed from home. In fact, due to light pollution, all but the brightest objects may be completely *invisible*. It's at this point that a new astronomer starts noticing advertisements for light pollution reduction (LPR) filters in the astronomy magazines. These would seem to be just the antidote for sodium-streetlight overdose. But, naturally, the new astronomer may be a little skeptical. Remove that horrible light pollution with the addition of a little filter? It just seems too good to be true. Do light pollution reduction filters really work?

Light pollution filters can work almost *magically* well. You will be amazed at the extent to which they can improve your views – if you understand how they work and what their limitations are. Holding a light pollution reduction filter up to your eye in front of a light reveals what looks to be a red or bluish colored piece of glass. It seems no different from any other filter used in astronomy or photography, maybe a little darker, but it certainly doesn't seem special in any way. In reality, though, LPR filters are made by a very different process from mere color filters. They consist of an optically flat piece of glass that has had numerous layers of reflective material deposited on it in a vacuum chamber. Each layer serves to *reflect* a different set of wavelengths of light.

Manufacturers choose these coatings depending on what types of light they want to be reflected by the filter and what wavelengths they want transmitted. This sounds complicated, but the filters work very simply. You install a filter on the end of the eyepiece that fits into your focuser (most modern eyepieces are threaded for filters). Light coming from the CAT's optical system hits this filter. The "bad" frequencies of light – light from sodium, incandescent, and mercury vapor lights – are reflected by the special coatings on the filter and aren't able to enter the eyepiece. The good light, light from a distant nebula for example, is not reflected and passes through the LPR filter and into your eyepiece and eye. This should make clear that, contrary to what some novice astronomers assume, an LPR filter does not make deep sky objects *brighter*. There is no light amplification involved. All the filter does is pass desired light from the objects. This has the effect of increasing the *contrast* between the deep sky wonder and the sky, making it stand out better.

This sounds very good. So good, in fact, the new astronomer may wonder why anybody would bother to travel to a dark site with these technological marvels available. LPR filters do work as designed, but they have some limitations. Chief among these is the fact that they do not work on *all* objects. Unfortunately, the light emitted by stars falls into the same range of wavelengths as that from earthly lights. This means light pollution filters are nearly useless on galaxies and star clusters. The light from the stars making up these objects is rejected along with earthly light pollution. If you decide to buy a light pollution filter, please be aware this device is really only effective on planetary and diffuse nebulae. This class of deep sky wonder can be helped tremendously by a filter, but if your main interest is in galaxies, an LPR filter might be a waste of money for you. LPR filters are not prohibitively expensive, but the best models *can* cost as much as a medium-priced eyepiece. LPR filters are being produced by a number of companies in the US and Europe, but the US companies Lumicon and Orion are the giants of the light pollution filter field.

If you do decide that an LPR filter can help your observing program, you may be surprised and confused by the number of different filters available for sale. Close examination of the specs of these many filters, however, reveals that they can all be classified in three categories depending on their *passbands*. Passband sounds like a forbidding and complicated high-tech word, but the concept is simple. A light pollution filter's passband refers to the range of frequencies of light allowed to *pass* through it. Filters are currently available with *wide* (mild or wide-band filters), *medium* or *normal* passbands (medium filters) and with *narrow* passbands (line filters). Each type is different and is suited to a different application. Most of the popular filters are available in either 1.25 inch or 2 inch sizes.

Mild Filters

Mild filters permit the widest range of light wavelengths to pass through them. Compare one of these to the other types of deep sky filter by holding it up to a lamp and it looks very light in comparison. It is letting most of the light pass through it. These LPRs are often referred to as mild filters because they have the smallest effect on deep sky objects. There *is* a contrast increase, because some of the bad light from the bright sky *is* being stopped, but it is less than that found in other types of filter. The US company Lumicon's Deep Sky filter is a mild filter, as is the other big American filter vendor, Orion's, Skyglow filter.

Why would anybody want to buy the *least effective* type of LPR filter? There are a couple of reasons. One strong point of the deep sky style filters is that they can be used in photography. The narrower-band filters would require punishingly long exposures when used with a camera. Another reason to choose a wide-band filter is, in the opinion of some observers, they *can* improve views of galaxies and star clusters. It is felt by some that these filters can darken the sky background enough to make these types of object look better without dimming their stars *too* much. Personally, I've never seen much of an improvement. In my eye, galaxies and clusters are dimmed enough by these filters to look no better than they did without the filter. In my opinion, you should choose a mild filter only in order to try deep sky photography from a light-polluted observing site.

Medium Filters

Medium filters, which are represented by Lumicon's UHC filter, are the workhorses of the LPR field. They are characterized by much narrower passbands than the mild filters. More bad light is blocked. This means they have a much more noticeable effect on nebulae than do the wide-bands. In fact, the first time you try one of these on The Great Orion Nebula (M42) from your light-polluted site, you'll be amazed. The improvement is truly dramatic. Trade-offs? The stars are much dimmer in these filters than they are in the wide-bands, making some objects a little less attractive than they were before. M42's nebulosity stands out *beautifully*, but the wondrously bright stars embedded within it are dimmed. If you only intend to buy one filter, a UHC, Ultrablock or equivalent is probably the one to get.

Line Filters

The passbands of line filters are narrower still. The best known of this class is Lumicon's OIII, which is a *very* high contrast filter. By the judicious application of many reflective layers, the manufacturer has produced a filter with a very narrow 10 nanometer passband centered on two Oxygen-III nebulae emission lines at 496nm and 501nm. What is this oxygen III? Why is it desirable? Oxygen III is the light of doubly ionized oxygen. It is often referred to in science texts as the *forbidden lines*. But what is important for you to know is that this is a wavelength of light that predominates in many nebulae, and especially planetary nebulae (the corpses of dead stars).

The OIII filter is truly amazing. My OIII, when used with my 8 inch SCT, has been able to turn the elusive Owl Nebula (M97) from an almost invisible smudge to an easily observed showpiece object (I can even make out the owl's *eyes*) in far less then perfect skies. If you want contrast, OIII's got it. It can improve the appearance of almost any nebula, and not just from bright suburban skies, but from the darkest of dark sites too. One of my fondest observing memories is of the normally somewhat elusive Bridal Veil Nebula in Cygnus as seen in an OIII filter from a very dark site. The OIII made this already interesting object into a thing of unending wonder. I spent at least an hour panning my Ultima C8 along its wispy and filligreed tendrils!

But there's always that piper to be paid. No, the OIII is not the filter for everyone. If you thought a UHC filter dimmed the stars, you haven't seen anything yet. The OIII actually *extinguishes* the dimmer field stars. And even the brighter ones take on a slightly reddish hue. There is no doubt this makes many fields a little less pretty. Another disappointment with this filter is that it does not work on every nebula. Most nebulae, diffuse and planetary, do enjoy a tremendous boost from the OIII, but some, those that lack OIII emission, aren't helped very much – if at all. Sadly, the greatest nebula in the northern skies, Orion's M42, is one of these. It always looks poorer to me with an OIII than without. Some observers think the OIII imparts *too much* contrast to objects, that the high-contrast views given by this filter look "cartoonish" and "not real." It's also a fact that because of its density, the OIII works best with telescopes of 8 inches aperture and up. It may dim views a little too much to make it a good choice for your C5 or 90 millimeter ETX. I've used and enjoyed the Lumicon OIII more than any other LPR filter.

Flashlights

Flashlights? Yes, you'll need a flashlight (torch) in the field in order to read star charts, set up and tear down the scope, select eyepieces and do many other chores that require you to be able to see what you're doing on a dark observing field. The question is, though, how to light things up while preserving hard-won night vision? Dark adapting your eyes is very important for the deep sky observer. In order to see the faintest objects your telescope is capable of delivering, at least half an hour of adaptation is required for your pupils to reach their greatest dilation. Expose your eyes to light and the process has to start *all over again*. What is needed is a light that is bright enough to allow you to read charts, but which won't do too much damage to your dark adaptation. For most amateurs this takes the form of a dim red flashlight.

The simplest observer's flashlight is a normal flashlight with a red filter added to it. One of these can be made easily by adding several layers of red paper or cellophane to a flashlight; an "astronomer's flashlight" of this type can also be purchased from astronomy dealers. The important consideration is that the red light be *dim* enough. A too-bright red light can do just as much damage to dark adaptation as a white one. When is a red light dim enough? Test it once you've allowed your eyes to become dark-adapted. Illuminate a chart or a white card with your new light and gaze at it for a minute or two. Then, shut off your flash and look up at the stars. My experience has been that just about all commercial red lights require the addition of more filter material. As shipped, they're just too bright.

But there are some astronomer's flashlights on the market that are perfect. These are not made by covering a normal incandescent light with filter material, though. Their light source is a red LED (light emitting diode). What makes these lights so superior? Their LEDs give off a much purer and redder light than a simple filtered flashlight can. The best units also include dimmer switches. This is a very handy thing to have on a flash. Star gazing from home, surrounded by lots of ambient light, your eyes are never going to reach full dark adaptation anyway, so you can safely turn up the brightness of the flashlight a bit for easier chart reading. At a very dark site, your pupils will be fully dilated, and you'll be able to adjust your light to a level where it won't have much effect on your eyes. Another benefit to these LED style observers' lights is that their batteries last a very long time due to the low current consumption of LEDs.

Star Charts and Atlases

A star atlas is a book of star charts that allows you to find the objects of your desire. While you were learning the

constellations or observing with binoculars you may have
been able to get by with the monthly star charts bound
into the astronomy magazines. But your powerful CAT
needs a much more detailed resource. A good, detailed
atlas is absolutely indispensable for *all* amateurs. But
which one? There are many collections of star charts for
sale at booksellers and from astronomy merchants.
Modern atlases, I've found, fall into three categories
depending upon the level of detail they offer.

Beginning Atlases

These sets of charts are also often referred to as the
Mag 6 atlases, since this is the faint magnitude limit most
of them feature. In the magnitude system, higher
numbers mean dimmer stars. Magnitude 6 stars are
close to the limit of what many people can see with their
naked eyes from dark sites. This chart group includes
some time honored books, including the well-known
Norton's Star Atlas. Some other more recent books in
this category include *The Mag 6 Star Atlas* (Tirion and
Dickinson), *The Cambridge Star Atlas* (Tirion) and *The
Bright Star Atlas* (Tirion). Wonderful works, every one of
them (Wil Tirion is a very gifted cartographer), but not
of much use to the average CAT owner, I'm afraid.

What's wrong with these books? Magnitude 6 is
simply not "deep" enough for our purposes. A mag
6 star atlas does not show enough stars to make finding
dim deep sky objects easy. 50 millimeter finders will
show *many* more stars than those depicted in these
books, making it hard to compare views between finder
and chart. Also, large areas of the sky in these atlases
may be sparsely populated with stars. A lack of
reference stars in areas like the Great Square of Pegasus
or Virgo's arms makes it incredibly difficult to track
down the deep sky wonders that reside in those places.
Another thing I don't like about the "mag sixers" is
their scale. The small scale of most of these works
makes them hard to read and less accurate than they
could be, even given their lack of stars.

Magnitude 8: *Sky Atlas 2000*

The middle ground is currently occupied by a single
work; one that has become the standard tool of the

modern amateur astronomer, *Sky Atlas 2000*. This
wonderful series of charts (again by master celestial
cartographer Wil Tirion) has been number one on the
amateur astronomy best seller list for many years. And
it's easy to see why. It's big, beautiful and deep. The
current editions of this work go down to magnitude 8.5,
nearly the limit of a 50 millimeter finder from many
locales. This ensures that every area of the sky is well
represented by many stars. In addition to stars, the atlas
features over 2,000 deep sky objects, a "best of the best"
which will keep even the most active observers
occupied for a very long time.

SA2000 is available in three editions. The Deluxe
Edition is spiral bound and features "black stars on
white sky" charts with color for the Milky Way, and
object symbols of various hues. This is the largest
version, with maps which fold out to a large 21 by
16 inches. Available for less money are the Field and
Desk Editions of *SA2000*. These are both composed of
unbound 18.5 by 13.5 inch charts. The difference
between these two is that the Field Edition has white
stars on a black sky and the Desk Edition has black
stars on a white background. Which of the three would
I choose? I find the "black star" Desk and Deluxe
Editions far easier to read in the field under a dim light
than I do the Field. The Deluxe has a nice binding, but
I've discovered that a few binder clips hold the Desk
and Field Editions together without a problem and
allow me to remove single charts for use if I wish. But
you can't go wrong with any of the *Sky Atlas 2000*
versions. Many amateurs find it's the only atlas they
ever need.

The Deepest

Sky Atlas 2000 served amateurs admirably well for a
number of years. But with the rise of the big
Dobsonian telescopes, the CCD cameras, and really
passionate deep sky observers, it soon became clear
that amateurs wanted even *more* detailed charts. The
first book to respond to this need was *Uranometria
2000* by Tirion, Rappaport, and Lovi. This amazing
volume presents 332,000 stars down to magnitude 9.5
and over 10,000 deep sky objects. This is done in two
hard-bound volumes composed of 259 9 by 12 inch
charts. Make no mistake about it: this atlas is not for

beginners. The small scale of the charts, 1.4 degrees per inch, means that the new user will get lost in a hurry. Except for the smallest constellations, each chart shows only a small part of each star figure's area of the sky. Even advanced amateurs will often be found using *SA2000* in conjunction with the *Uranometria* to help them navigate.

The next highly detailed atlas to appear was the product of the hard work of two Australian amateurs, David Herald and Peter Bobroff. Their atlas, *The Herald–Bobroff Astroatlas* is a completely new paradigm for the star chart. The heart of the atlas is a series of large 11.5 by 16 inch sky maps that show stars down to magnitude 9. In addition to the stars, the atlas also features 13,000 deep sky objects – enough to keep even the most sanguine deep sky fanatic occupied for a very long time. But this is only *part* of the *HB*. The atlas also contains several other series of charts. There's a set that goes down to only about magnitude 6. This is like having an entire *Norton's* included in the same book as the more detailed charts. But it gets better. The magnitude 6 charts are marked with letters and numbers that refer you to the magnitude 9 charts.

Let's say you're browsing through the magnitude 6 atlas section. You find a particularly interesting area, and would like more detail. You note that the legend "C6" is overlaid on this area of the chart. You turn to the magnitude 9 series chart "C6" and find the same area but zoomed in, showing much more detail. This feature of *Herald–Bobroff* means you can leave your poor old *SA2000* at home and only lug one atlas into the field. Need more depth? Well, *HB* has still more pages of maps. Turning to the back of the book, you'll find charts for selected areas of the sky down to an amazing magnitude 15!

Herald–Bobroff still not deep enough for you? Well, how does an atlas containing over 1,000,000 stars down to magnitude 11 strike you? This is Sky Publishing's *Millennium Star Atlas* (Sinnott and Perryman). In addition to its multitudinous stars, *MSA*'s three volumes of 1,548 9 by 13 inch charts offer a generous portion of galaxies, clusters and nebulae – more than 10,000 deep sky objects are represented. And there's not just a lot here, it's all in the *right places*. This book was plotted using the star position data obtained by the European Space Agency's groundbreaking *Hipparcos* satellite. This surely must be the ultimate – for a while, anyway – in printed atlases! Like *Uranometria*, this book is really aimed at the advanced amateur (or even professional). There's so much here in such fine detail that I'm afraid the average beginner is going to find the book completely indecipherable. But for the advanced deep sky observer this is awesome!

Observing Tables

A table is a telescope accessory? You bet it is. You'll need somewhere to put all these accessories I've urged you to buy. In addition to space for your star atlas to lie flat, you'll need room for an eyepiece case, a flashlight, and all the other small items you'll want to keep close at hand during an evening's observing run. Any portable table will do, but I've found that the best compromise where size and weight are concerned is a plain, old folding card table. This is easy enough to transport in your auto with the legs folded up, and provides enough room for a decent array of accessories (though I now have enough astro-junk that I'm considering adding a *second* card table to my setup). Some astronomy merchants and camping supply stores sell card table sized tables that fold and roll up into a package only a few inches in diameter and a few feet long. These are fairly steady and may be a good choice if you frequently travel to distant observing sites.

Now that you're loaded down with accessories, only one thing remains to be done: get out in the field with that beautiful new CAT and start using it! The next chapter will take you through a complete initial observing/checkout run with a new telescope.

Chapter 7

From That Very First Night: Field Setup, Checkout and Observing

The time has finally come for you to stop being just a telescope *buyer* and become a telescope *user*. This chapter will cover what you need to know to set your CAT up in the field and begin observing.

To begin, you'll need to find a spot outside for the telescope. On this first outing with the new scope, I'd suggest you stick close to home, even if your skies are badly light-polluted. If this is your first CAT, you'll likely spend as much time going over the instruction manual and fiddling with your scope's fittings as you will looking through the eyepiece. This is one occasion where you might find it *helpful* to be able to use a white light occasionally if you run into problems. Start flashing a white light around at your club's dark observing site and you'll soon be *persona non grata*.

Where exactly do you set up your SCT in your backyard? Look for an open space, especially one with a nice view of the northern and eastern skies. You'll need to be able to see clearly to the north in order to polar align your scope, and you may find that, like me, you tend to gravitate toward the objects rising in the east more so than those disappearing in the west. Try your best to find a spot where you'll be shielded from neighbors' porch lights and security lights and from nearby streetlights. Unfortunately, if you live in a typical suburban location, this may be difficult – lights will be all around you. On this night, just do the best you can.

Ideally, you should take your CAT outside at *least* half an hour before you plan to begin observing. This will allow it to cool down, to adjust to outside temperature. When you first bring your telescope from a warm indoor environment into cooler evening air, your mirror will begin to contract and will lose its perfect shape, delivering images that are far from the best it can produce until it reaches ambient temperature and settles down. You'll also be looking through hot air "tube currents" on the inside of your OTA, and these will really mess up images, which will shimmer and waver as if you are looking at them from underwater, until your OTA cools and the tube currents disappear.

Some observers will tell you that even half an hour of cooling is not nearly enough if you're interested in optimum images or if the temperature differential between outside and inside is great. In these circumstances, you may need to allow at least 1–2 hours for an 8 inch SCT to cool down. Maksutov–Cassegrains with their big "salad bowl" corrector plates may take even longer to cool off than SCTs, and SCTs larger than 8 inches require more time than smaller models. But cool own time needn't be a big problem. My C8 goes outside as soon as I return home from work in the afternoon, and by the time darkness comes all is well. Just don't place the telescope where sunlight can fall on it, or your cool-down time will be dramatically extended.

Position the scope's tripod and wedge combination in the chosen observing location. Spread the legs out to their full extent and tighten the leg spreader into place. Remember, not too tight. Orient the tripod so the flat surface of the wedge, the surface where the scope's drivebase will be mounted, is pointed north. You don't have to be precise at this time, but the closer you get to true north, the less moving the tripod around you'll have to do when the telescope is mounted on the wedge. Use a compass if you're not sure which way north is on your property. To save time in polar aligning, you'll also want to adjust the tilt plate to your latitude angle if you haven't done so already. When you've got the plate at an angle where the pointer on its scale matches your latitude, tighten it down again.

To mount the telescope on the wedge (or the tripod alone if you're using a computerized alt-az mode scope), follow the very same procedures in Chapter 5 you used when you set the CAT up in your living room for its initial checkout. If at all possible, carry the telescope in its case or box outside and place it near the tripod. Slide it onto the wedge's tilt plate just as before, so that it is "hanging" by the single top bolt you've threaded in. Tighten this bolt a little to make sure the scope is secure on the wedge and insert the other two bolts through the plate and into the drivebase. Tighten all three bolts securely.

While you're waiting for darkness and for your scope to cool down, install your accessories. If the finder was removed for storage, attach it to the OTA. Remove the rear port cap and thread on the visual back. Install the star diagonal too. Insert a low-powered ocular into the diagonal. Some people will advise you to wait to affix the star diagonal/eyepiece, leaving the rear cell uncapped to speed cool-down time. In my experience, however, this doesn't help much, and could allow foreign objects to enter your tube. For now, leave the lens cap on the corrector end of your scope for protection.

If your telescope requires an external power source, a battery or an AC connection, set that up now. Locate your hand control and plug it into the control panel, making sure the power to the drive is still off before you do so. Once the hand controller is plugged in, you may want to turn the scope drive on momentarily to check that there's power. If not, check the connections and the status of the battery. If you wish, you can also install any powered accessories you may have at this time – focus motors, declination motors, digital setting circles. Actually, you might as well leave all these extras off tonight. You won't be using them and they may get in the way as you're doing this initial field testing.

With darkness falling, it's time to prepare your dew-fighting tools. If your only anti-dew provision is a dew shield, you can set it aside until it's time to start observing. If you use one of the little blow dryer dew zappers, connect it to a storage battery or other power source. If you've invested in a corrector plate heater, install it now, following the instructions that accompanied it. Some people attach the corrector heater just behind the corrector plate on the tube of the scope, but I've had the best results by installing it right over the corrector assembly on the end of the tube. If you own a dew shield, remove the lens cap now, slide the dew shield over the end of the scope, and fasten the heater *over* the dew shield's base at the end of the OTA. If you have a dew shield made of foam or some other thick material, it probably is advisable to install the corrector heater just behind the corrector assembly on the scope's bare tube. Turn the dew heater on as soon as it's in place. This will allow the corrector to warm, warding off dew as the Sun sinks and the temperature drops. The telescope is now ready for the evening's observing run But are *you?*

Even if you're only observing from the backyard, you need to take steps to prepare yourself for the evening ahead. This consists of keeping warm and keeping insects away, since cold and bug bites will shut down an observing session quickly. Even in the summer, always dress more warmly than you think necessary. During the hottest times of the year it is still very easy to get chilled when you're standing almost motionless under the night sky. If you're observing at home, you can always run inside and warm up, but this quickly becomes annoying and will probably cause you to give up before you've seen everything you'd hoped to see. Your climate will dictate exactly how you dress, but wherever you are, dress in layers. Layers of clothing rather than one heavy coat or sweater will help your body retain heat. Much of the body's heat loss is through the feet and head, so take special care to insulate them. On a fairly mild evening, this may be as simple as putting a small rubber-backed bathroom rug next to your scope to stand on to isolate your feet from the cold ground (works surprisingly well). On bitter nights, you may need special boots designed for harsh conditions. Cover your head with a hat of some kind, of course.

You should also try to keep your *insides* warm on bitter nights. On really bitter nights, a thermos of hot drink will be most welcome in the wee hours. According to the experts, the best beverages are caffeine-free, like hot cider. Apparently, caffeine can have a detrimental effect on your ability to remain warm. I have always found, however, that coffee stands me in good stead. Don't drink alcoholic beverages. They dilate blood vessels in the skin, causing you to actually grow colder much more quickly than normal; even though you may feel that that shot of brandy or whiskey has warmed you up. Alcohol also seems to have a bad effect on dark adaptation. Save the booze for *after* the observing run.

Insects? Depending on your climate, mosquitoes can be a real problem. If you live in subtropical or tropical areas, these insects can easily put an end to an observing run if you don't take some steps to deal with them. You'll hear many suggestions on how to keep the mosquitoes away, and you'll find store shelves stocked with many nostrums guaranteed to keep the little suckers from biting, but there's only *one thing* that really works: DEET. DEET is the abbreviation for a chemical, *n, n-diethyl-m-toluamide*, which must be present in your insect repellent for it to be really effective. If you *must* keep insects away, make sure that DEET is an ingredient in the product you purchase. DEET-based repellents work great, but they can have a solvent effect, so be sure to keep them away from optics and plastic/rubber telescope parts.

Finder Alignment

Telescope's setup. Accessories are in order. You've dealt with dew and insects. What now? A good place to start, before attempting polar alignment, is finder scope alignment. If this is a new telescope, or if you normally remove the finder to return the scope to its case, you should check your finder's alignment. By "alignment" I mean making the finder scope's field of view coincident with that of the main telescope. The finder must be adjusted so that whatever is in its cross-hairs is in the middle of the SCT's eyepiece field of view. To do this, insert your lowest powered ocular in the main scope, uncap the finder, and select an earthly alignment tool. Choose a telephone pole, a tree or some other object distant enough for your main scope to focus on and far enough away to eliminate parallax problems between the main scope and the finder. Undo the RA and declination locks on your CAT at least partway, enough so that the scope can be moved freely. If you haven't done so already, remove the lens cap from the main scope's corrector.

Carefully move the telescope in RA and declination in the direction of your chosen landmark. As you approach the area of the target, look through your finder, and continue to slowly move the OTA until you have your landmark in the cross-hairs. When the target is centered, lock both telescope axes. If the image in the finder is blurry, adjust the focus until it's sharp as per the directions that came with it. Look for an easily identifiable marking on your target to help with your alignment – a particular insulator on a telephone pole, a bird's nest in a tree, or any other readily distinguished feature. Center this feature in the finder cross-hairs using your scope's slow-motion controls (remember to at least *partially* release the RA lock before turning the slow-motion control on this axis) and take a look through the main telescope. If all you see is a blur, turn the focus control until the image becomes sharp and clear. If it doesn't seem to be coming into focus, turn the control in the opposite direction until it does so. If you can't achieve focus, it's possible that your landmark is just too close to the telescope – choose another.

Is the chosen feature on your target in the main eyepiece? Chances are that it's not. Turn the declination and RA slow-motion controls until the reference point is centered in the main eyepiece. You may find you have to hunt around a bit before you can get your CAT lined up on the right spot. You'll also no doubt notice that the difference between the view in the finder (an inverted but mirror correct image) and the main telescope (right side up but mirror-reversed image) is a bit confusing at first.

What if the telescope does not appear to be pointed anywhere near your chosen landmark? In that case, the most expedient method of getting it aimed is to just sight along the tube and move the scope carefully (undoing the locks a bit) in larger increments until it's centered in the right place. This may take some trial and error work, but keep at it – having an accurately aligned finder is critical.

You should now have your alignment reference point centered in the main scope's eyepiece. Lock both RA and declination locks again and take a look through the finder. If you had to move the telescope at all in order to place the target object in its field, it will no longer be exactly centered in the finder scope's cross-hairs. Return the object to the cross-hairs by adjusting the finder's *mounting*. Most finder scopes sold for use with SCTs are held in "ring" mounts. Either one or both of the rings that hold the finder in place will have screws (usually three apiece) that can be turned to move the finder within the rings in order to adjust its aim. Start with the forward ring. Loosen the adjustment screws' locknuts (if there are any) so all three screws can turn freely. Looking through the finder, decide which way you need to move the image in order to center it in the cross-hairs.

Because the image is upside down in most finders, choose the screw opposite this direction and experimentally loosen it. If the image moves in the right direction when you loosen the screw, keep going until the image is dead center in the junction of the cross-hairs. You will also need to tighten the two screws opposite the one you're adjusting in order to keep the image moving the right way and to secure the finder when you're done. If the image didn't move in the right direction when you loosened the screw, try tightening it slightly (you may have to loosen its two "opposite numbers" in order to do this). If this moves the image the right way continue until it is centered. Try to place whatever marker is in the center of your main scope's field exactly in the cross-hairs.

When you're satisfied with the adjustments you've made to the finder, remove your low power eyepiece and place a medium-powered one in the diagonal (about 100–150×). Focus up and, if necessary, move the slow-motion controls to recenter your marker – insulator, bird, sign, whatever – in the field. Tighten the RA and declination locks and look through the finder again. Is your marker still dead in the center of the cross-hairs? If not, adjust the finder screws until it is as close to the junction of the finder's cross-hairs as possible again. If for some reason the screws on the forward ring don't allow you enough "travel" to precisely center the object in the finder, you may use the rear ring's alignment screws to finish the job.

When you've finished this task, the stars should definitely be winking on. Choose a bright and prominent one to finish the alignment of your finder. Remove the medium-power eyepiece and replace it with your lowest-power ocular. Loosen the scope locks as you did previously, and slowly and gently, with one hand near the front of the tube and one on the rear cell, move the OTA in the direction of a chosen bright star. When you start getting close, put your eye to the finder eyepiece and move it into the cross-hairs just as you did with the terrestrial landmark. Use the slow-motion controls if necessary to place the star exactly in the finder cross-hairs. When it's there, gently engage the RA and declination locks. If the earthly object you used for finder alignment was not at "infinity," you may have to refocus the finder scope until the star is as sharp and point-like as possible.

Look through the main scope. If the star looks like a doughnut with a dark center, you need to focus. Turn the SCT focuser until the star is as small as possible. When it's sharp, *just enjoy the sight of your CAT's first star for a moment!* When you get over that wonderful feeling that comes with First Light for your new telescope and finish yelling for your family to "come look at the beautiful star," examine the position of the stellar image. To finish our finder alignment, use the slow-motion controls to place the star as close to the center of the field as you can. Even if it was in the center when you first looked through the eyepiece, it will be "off" now, because we haven't aligned the mount or turned on the drive – the rotation of the Earth is carrying the star out of the field of view. Look thorough the finder to see if the star is close to the center of the cross-hairs. If not, adjust the finder alignment screws until it is. Continue adjusting the finder and recentering the main scope on the star until you're satisfied.

Rough Polar Alignment

When the finder scope is aligned, you can begin the polar alignment of your telescope mounting. Polar alignment is the process of pointing the right ascension axis of your telescope at the North (or South) Celestial Pole (see Figure 7.1). Once this is done, your equatorial mount will be able to track the stars – to counter the Earth's rotation – allowing you to keep objects in the field of view for as along as you wish without having to move the telescope by hand.

A polar alignment accurate enough for astrophotography is an involved and lengthy process (it will be described in Chapter 10), but an alignment good enough for tonight's plan – casual visual observing – is quick and easy. The following instructions apply to fork-mounts, but can also be used, with few alterations, to align a German mount. The main difference in aligning a GEM is that you usually look through the telescope's hollow RA shaft (often through a special borescope) to aim the mount's RA axis at the pole (or here, simply at Polaris). Consult your mount maker's instructions for further details.

To begin polar alignment of your fork-mount SCT, set the telescope's declination to 90 degrees using the declination setting circle for reference. Undo the dec lock and turn the scope until it looks like the illustration in Figure 7.1 and the declination setting circle reads 90 degrees under its pointer. Lock the declination lock and unlock the RA lock. Turn the

Figure 7.1. Set your SCT at a declination of 90 degrees to begin polar alignment.

scope in RA until the fork is level with your finder on top, also as in Figure 7.1 and re-lock the RA axis. From now until you finish polar alignment, leave both locks engaged. When the telescope's position has to be changed, you'll change it by *using the wedge's altitude (up/down) and azimuth (side to side) motions.* Your goal, if you live in the northern hemisphere, is to move the telescope up/down and right/left until Polaris, the North Star, appears in the finder telescope and in the field of the main scope.

In order to do this, you obviously have to know where Polaris is. Polaris, a magnitude 2 yellowish star, is a reasonably bright sparkler that will always lie due north in the sky. While it is prominent, it is not overwhelming, and if it were not for its position marking the North Celestial Pole (actually it's about a degree from the true pole), it would be an interesting but not remarkable star.

Since Polaris is the end star in the Little Dipper's handle, you can use this constellation to locate it easily if light pollution has not made this dim group of stars invisible from your area. Another way to locate Polaris is by its altitude and azimuth. Using a compass, if you don't know exactly which direction north is, look north. Polaris will be in this direction and at an altitude *equal to your latitude*. If your latitude is 40 degrees north, Polaris will be due north and 40 degrees above the horizon. It's easy to judge angular distance in the sky, by the way. You have a built in measuring tool, your hand. Close your fist and hold it at arm's length against the sky. The distance from the end of your thumb to the other side of your clenched fist is 10 degrees. Four fist-widths up will be 40 degrees. You should notice a bright, slightly yellowish star in this approximate area (because of magnetic deviation, Polaris may not be in the exact direction of north as indicated by your compass).

If you've carefully placed your wedge and tripod so that the tilt plate is facing north and have adjusted it so that it's at the proper angle for your latitude, it's probable that Polaris will be somewhere in the field of your finder when you look through it. If Polaris is *not* visible when you look in the finder, you'll have to sight along the OTA and make side to side and up and down motions until it is in the field. To do this, loosen the altitude and azimuth adjustments of your wedge in accordance with the manual. Be careful, if your wedge does not incorporate a latitude fine adjuster, not to loosen the tilt plate too much. With the weight of telescope on it, it will tend to fall over. Just loosen the bolts enough to allow you to change its angle using some force. If you find it difficult to adjust the altitude of your wedge with the scope in place, you can try extending or collapsing on of your tripod's legs. Just don't do this to the point where your telescope is so off balance that it could collapse. Also be careful in loosening bolts or knobs that allow side to side movement of the wedge. Don't back them off so much that the wedge and scope are in danger of falling off the tripod! Once you can see Polaris in the finder, you must move it squarely into the center of the cross-hairs. Remember, use your wedge altitude and azimuth adjusters – don't touch the RA and declination controls *at all*.

Once Polaris is in the middle of the finder field, take a look through the main telescope's eyepiece. If you've aligned your finder carefully, Polaris should be in the center of the low-power eyepiece field. Not sure if you've got the right star? You can *easily* tell whether you're looking at Polaris or not. This second magnitude star will appear bright, yellowish, and, perhaps surprisingly, will turn out not to be a *single* star. Even at 80×, you'll see that Polaris is accompanied by a dimmer companion star about 18.5 arc seconds away. Polaris is a *binary star*. The single star you see with your naked eye or with low magnifications is actually a pair of stars in mutual orbit around a common center.

If your wedge is equipped with fine adjusters, you can use them to place Polaris *exactly* in the center of your main scope's field. If your wedge *doesn't* have a means of fine-tuning, don't worry about centering the star precisely in the main scope. For non-critical visual observing, placing Polaris in the field of the *finder* is sufficient. When you're satisfied that you have Polaris in view in the main telescope, tighten down the wedge's tilt plate and azimuth bolts. Do this carefully so as not to disturb the scope's position.

Southern Hemisphere Polar Alignment

What if you live in the southern hemisphere? You certainly can't polar align on Polaris – it's invisible. You will have to align your scope on the *South* Celestial Pole. Polar alignment in the southern hemisphere is a little more complicated than it is north of the equator. This is because, unfortunately, there's no bright star to mark the location of the South Celestial Pole. The South Celestial Pole lurks among the relatively dim stars of the far southern and somewhat obscure constellation, Octans, the Octant. The Southern Pole Star, the star that is currently closest to the South Celestial Pole, is Sigma Octanis. This South Pole Star is pretty dim at magnitude 5.4. It *is* just about as well *placed* as Polaris, however, being approximately 1 degree from the South Celestial Pole at this time.

Sigma Octanis should be visible through a 30 millimeter finder, but if you're not very familiar with this area of the sky, you may need to consult a good star atlas to make sure you're aligning on the correct star. Once you've got the right star centered, be sure to set your telescope's drive system for southern hemisphere operation.

Many beginners worry needlessly about polar alignment. The foregoing easy process of pointing the RA axis at the pole (or Pole Star) is *all* that's required for most tasks. What you've done is *completely sufficient* for visual observing. The occasional adjustments you'll have to make with the declination or RA slow motions to correct for misalignment drift won't get in the way of your observing.

Power Up and Drive Checkout

If you haven't done so already, turn on your drive and let's get the telescope tracking. If you've already performed the drive-check steps outlined in the indoor setup section of the last chapter, you've already made sure that your drive runs and that the additional drive rates seem to work correctly. You now need to determine whether the drive is tracking. Select sidereal rate if you have a choice of drive speeds, and, using your finder telescope, point the CAT at a bright star. Any bright star will do, but one conveniently located in the east will be easy to work with. When this star is centered in the cross-hairs of the finder, move your eye to the main scope. Start out using your lowest powered eyepiece. If the chosen star is not centered, use your hand controller buttons to touch up its position in RA (and declination, too if you have a declination motor). Using the hand control, if you have one, will give you a good feel for your drive's guiding speed.

Observe your chosen star for a few minutes. Does it move in the field? If you've done a good polar alignment, you should notice very little change in the star's position. Remove the low-power eyepiece and insert one that yields a magnification of about 150×. This will make any mis-tracking by your drive more evident. What should you see at high power? You will now undoubtedly notice a little drift. This is normal. If you've aligned by pointing the RA axis at Polaris, you are not *exactly* on the celestial pole. The direction of drift will be dependent on your misalignment – whether you are north/south or east/west of the true Celestial Pole. When a planet or deep sky object drifts toward the field edge, no big deal. Use your hand controller or manual slow-motion controls to recenter it.

But what if a star or other object leaves the field very quickly? In just a couple of minutes or less? Make sure you've *really* got power to the drive. You can usually tell if the motor's running by listening carefully and by turning it off and on if necessary to note the difference. If the motor doesn't seem to be working, check your power connections and the status of your battery, if you're using one. If the motor is running, *check to see that the RA lock is engaged*. If the lever is not at least partially locked down (at *least* halfway), the motor will not move the telescope. Many beginners forget this in the course of locking and unlocking the RA axis to find and center objects. If the lock is partially engaged, lock it down all the way and see what happens.

Another reason for poor drive tracking? Did you *really* align the scope on *Polaris?* If you're new to all this, it's quite easy to pick the wrong star for your polar alignment. Your author has even done this a time or two, usually when he's under a dark, desert sky for the first time in many months and is confused by the hordes of now-visible stars. If you suspect you've aligned on the wrong star, return the scope to 90 degrees declination and take another look at "Polaris." Remember, it should appear as a golden-yellow star with a close, dimmer companion. If you misidentified the Pole Star, you'll need to redo your polar alignment – being very far off will cause extremely poor tracking. If none of these checks resolves the problem, see the troubleshooting section of this book in Chapter 8 and your user's manual for further assistance.

When you're satisfied your drive is operating well, you can test its other features, its various slewing speeds and drive rates. Try these out, zipping around until you find a speed you like for fine centering and one that seems to work well for more coarse positioning of the telescope. When you release the high-speed button(s) your telescope should resume tracking normally. Please note that some modern telescope drives will move the scope a small distance in the *opposite direction* after you finish a high-speed slew. This is to help take up backlash in the gears and does *not* indicate a problem of any kind. You should not notice this behavior after using the slowest guiding rate, though.

Aligning Goto Scopes

If you're the owner of a goto CAT – an Ultima 2000, LX-200, Nexstar or ETX – you may feel a bit left out, since none of the polar alignment/drive checkout procedures really apply to you. You have your own job to do, though, which consists of getting the telescope and its computer aligned and tracking in (usually) alt-azimuth mode. Following the installation of your scope on its tripod in the same manner as when you assembled it inside, you'll get your high-tech marvel going by aiming it at one or two known stars. The exact procedure you follow will depend on the model of telescope you have. In some cases, the telescope computer will choose the alignment stars for you. With other units, you may be allowed to make your own choices. If you are allowed to choose your own alignment stars, follow directions in your manual as to how far apart they should be in the sky for best results. You should also carefully observe any instructions concerning the way your telescope's base and tripod should be aligned. With some models you may be required to face the telescope's control panel, for example, in a certain direction.

With an alt-azimuth mode telescope, a poor alignment will not only result in less accurate goto operation but also in decreased tracking accuracy. It is advisable to use a reasonably high power eyepiece to center stars during the alignment. If you're equipped with an illuminated cross-hair eyepiece, this will go a long way toward helping you do accurate alignments. About 150× is a good magnification. Once your alt-azimuth telescope is set up and aligned, check its tracking accuracy and various slewing and tracking rates in the same manner we did with the fork-mounted equatorial telescopes.

Beginning to Observe: The Moon and Planets

It's *finally* time to use this telescope, the one you've agonized over for months during the process of choosing it, ordering it and checking it for proper operation. And what better subject for a CAT's first light can there be than the Earth's faithful companion, Luna? Lucky indeed is the amateur who's fortunate enough to find a nice crescent or gibbous Moon hanging in the sky on that first evening (a full moon doesn't reveal many details; the Sun angle is too high for lunar features to stand out in good relief). Many advanced amateurs dismiss the Moon. It's too bright, too "easy," and its silvery light spoils their views of the dim galaxies and other distant wonders they pursue. I've done lots of deep sky observing and enjoy a faint nebula as much as the next amateur, but, even after 35 years, I still find the Moon endlessly fascinating.

Are you ready for a voyage of discovery? Let's point that new SCT at the wonderful Moon. As you've done before, partially loosen the RA and declination locks, place one hand near the front of the tube and one hand on the rear and gently guide the scope toward Luna. When you get in the general neighborhood, put your eye to the finder eyepiece and guide the Moon into the center of the cross-hairs. Gently lock the locks and insert your low power eyepiece into the star diagonal. You'll probably be half blinded at first. The Moon is bright and an 8 inch or larger scope at low power does deliver a lot of light. Many beginners are struck by how bright Luna is through their telescopes and wonder if this intense image might be *harmful*. Don't be concerned, though. The light of the magnitude −12.7 Moon is not anywhere near intense enough to damage your eyesight.

Once you get over the shock of how bright our satellite world is, you'll be amazed at the *incredible* wealth of detail your telescope is capable of revealing on this ancient relic. If the image of the Moon doesn't appear to be crystal clear, adjust your SCT's focus control until things are as sharp as possible. Focus by looking at the craters and mountains that stand out in stark relief along the Moon's day/night terminator. And then just look for a while. Use your slow-motion controls or hand controller to adjust the telescope's position as necessary.

When you've had your fill of Luna with your low-power eyepiece, switch to a shorter focal length one. You'll notice that a higher magnification eyepiece makes the image dimmer. But once you adjust to the change in brightness, you'll find that you're able to detect more details than before. Not only do you see craters, you see *craters within craters*, and other microfine details. When I'm observing the Moon, I keep increasing my power until I can't see *any* additional detail. At some point more magnification will not show me anything more. The image will begin getting fuzzier as well as bigger. This will be due to either unsteadiness in the Earth's atmosphere or the size, quality and alignment (collimation) of your optical system. Magnification that doesn't produce an increase in visible details is often referred to as "empty magnification" by amateurs.

This is a good time to discuss magnification. How much is enough? What's too much? For the Moon or any other subject? The very freshest novices usually think more power is always better. The 60 millimeter refractor they found in the department store with a box emblazoned with 750× must surely be better than the SCT in the astronomy catalog with an eyepiece that only gives 80×. If the beginner continues growing and learning in astronomy she or he soon learns high power isn't the measure of a telescope's usefulness, *aperture* is. At this point, new observers with new scopes tend to shy away from higher magnifications, tending to think high power is "bad" and that using it reveals them as novices.

High magnifications do have their places, though, and especially in lunar or planetary observing. For seeing detail, fine detail, on the Moon, I use powers of 200–250× and often much more. When searching for details in the floors of craters, I've pushed my C8 to 600× with good results. Many amateurs will tell you that 600× is too high for an 8 inch telescope, citing the common wisdom that the maximum useful magnification of your telescope is about 60× per inch of aperture (meaning that I shouldn't be using more than 480× on my 8 inch). This formula is a good *broad guide* for what will work with a scope magnification-wise. But it doesn't *always* prove true. If you live in an area blessed with steady atmospheric seeing, a telescope with good optics and have collimated those optics precisely, you may often be able to exceed the "limit" on a bright object like the Moon. On the other hand, if your area suffers from poor seeing, or your CAT hasn't had its optics aligned in a while, you may not be able to come close to the 60× per inch figure.

Lunar Features

Craters

Even a casual glance at Luna in the telescope reveals that the lunar landscape can be divided into two general types of terrain, highlands covered with the ring-shaped formations called craters and the relatively smooth areas formed by the mare, the lunar "seas." The highland area, and particularly the southwest quadrant, is a veritable paradise for the Moon lover, as it is composed of unending hordes of shoulder to shoulder craters of all types, shapes and sizes.

These seemingly innumerable craters are the product of aeons of bombardment of the airless Moon by solar system debris, comets and asteroids. Lunar craters range in size from the tiny, less than an inch across, to great dishes hundreds of miles in diameter. One of the first things the beginning observer wants to know is, "What's the smallest crater I can see?" A well-collimated 8 inch SCT under good conditions of atmospheric seeing can reveal craters ½ mile across. The shadows created by the crater walls at lunar sunset and sunrise mean that your SCT can distinguish craters which would normally be too small for it to make out, that would be beneath the *theoretical resolution* for a particular size of SCT. A 12 inch CAT might be able to distinguish craters *somewhat* smaller even than ½ mile in diameter, but generally bigger telescopes won't do too much better in this regard than an 8 inch, because the Earth's always unsteady atmosphere limits the angular resolution of even large telescopes.

The Maria

After the cratered highlands, the most noticeable features of Diana's silvery countenance are the maria. It's been pretty obvious for hundreds of years that the dark areas on the Moon's face visible to the naked eye aren't really seas. Even a tiny telescope or a pair of binoculars reveals them for what they are: incredibly huge plains surfaced in a dark material. A 5 inch SCT or even a 90 millimeter MCT will show the maria are speckled with craters, crater ejecta, mountains and other solid features.

At first, the Moon's plains seem much less interesting than the highlands, but these areas have their own features of interest. One thing you'll note right away is that the dark, dried lava material which covers these areas is not of a uniform color, but can vary over a fairly wide range from sea to sea and even across the larger maria. At moderate magnifications, you'll be able to see that the supposedly smooth maria are not even close to being really smooth or featureless. Like the highlands, these areas are also home to a great many craters, just in less profusion. Some of the most dramatic craters, in fact, are visible in the midst of the seas, magnificent edifices like Copernicus and Kepler.

Other Features

Even a 90 millimeter ETX will show that craters and mare aren't all there is to see on the face of Earth's sister. Systems of rilles, cracks in the lunar surface, are visible in many locations, stretching on for hundreds of miles and forming intricate networks. There are also valleys, like the magnificent and imposing Alpine Valley near the great dark floored crater Plato, and scarps, places where the lunar surface has been elevated in linear fashion, forming great cliffs. The Straight Wall scarp, visible in an ETX or C5 with ease as a razor-thin, black line is an example of this type of feature. Less obvious than any of these are lunar domes, gentle swellings of the surface. These strange features may have been created by volcanic activity and are almost impossible to detect except under the lowest possible Sun angles.

Tips for Lunar Observing

Accessories

There are a few additional items you may want to add to your growing collection of accessories to make your lunar exploring more productive. The most important item is a map of the Moon or a lunar atlas. As you'll see the first time you turn your telescope to the Moon's face, it is a maze of confusing details. A map is vital for making sense out of the Moon's landscape. In the beginning, a simple large-scale map such as one of those found in many general observing guides will be enough. These identify the major features, the craters, seas and a few of the other easily seen details. If you really become interested in the Moon, you'll need something more detailed. A very good small-scale map is offered by Sky Publishing, Inc. This is the *Lunar Quadrant Map*, a set of four large sheets, each of which is a detailed quarter of the Moon. These charts depict most of the features easily seen in a small telescope. On a more advanced level there is a book that has become the standard work for committed Moon watchers, Antonin Rukl's *Atlas of the Moon*. This is a true atlas, with the Moon's visible disk portrayed in considerable detail in beautiful airbrushed maps. If you're using a star diagonal, the view in your SCT will be mirror reversed with regards to your chart. You can deal with this by copying charts onto transparency film with a Xerox machine and flipping them over for a correct view.

The Moon is, as we've said, not dangerously bright. It can, however, be bright enough that it can be difficult to make out small details because of the glare. Many observers turn to Moon filters to dim the Moon down a little. These are neutral density filters, filters that are not colored but only serve to reduce the intensity of the light. Like most other astronomical filters, they screw into the end of the eyepiece that goes into your star diagonal or visual back.

I don't really recommend Moon filters for lunar observing, though. In my experience they reduce light too much for many telescopes, even when used with low-power eyepieces. They also don't do anything to *enhance* the appearance of the Moon's features. All they do is attenuate the light. A slightly better choice for someone who wants to reduce Luna's glare is a polarizing filter. A filter of this type is actually an assembly containing *two* polarizing filters. They are arranged so that one of these filters can be rotated with respect to the other. Because of the special nature of polarizing material, rotating one layer causes less and less light to be transmitted through the pair. Turn far enough, and the filter lightens again. A polarizer allows you to "dial in" the darkness of your Moon filter.

In my opinion, however, the best solution for dimming the Moon is a colored eyepiece filter. These are available from scope merchants in a wide variety of colors and in 0.965, 1.25, and 2 inch sizes and can work just as well as a Moon filter or a polarizing filter to dim the Moon's glare. Some colors can also enhance surface detail. An 80A blue filter (the number and letter associated with a filter is its *wratten type*, a standard system for indicating filter color and density), for example, both reduces glare and enhances small details. A #15 yellow can make ray systems and rilles stand out better. A good variety of colors (which can be used in combination, too) will add to your lunar program. It's fun to try various shades alone and together to see what the results might be.

Many modern SCTs have a lunar drive rate selection. The Moon moves at a speed just slightly faster than the sidereal rate. Switching over to the "lunar" position will allow your telescope to track the Moon exactly. You can use this drive speed when doing lunar observing if your scope features it, but in practice the difference in sidereal and lunar rates is small enough that I usually don't bother with it. Even a plain old, AC synchronous motor SCT will track Luna very well assuming it has been accurately polar aligned.

Planetary Observing

The Moon isn't the only near sky object available to the CAT owner. The Sun's family of planets offers countless hours of observing enjoyment and challenges for even the smallest CATs. One advantage of the shallow sky is that, except for dim Pluto, the planets are not affected at all by light pollution. You can get just as good a look at Jupiter or Saturn or Mars from an urban site as from a dark country location.

A Brief Tour of the Solar System

The Sun

The ins and outs of solar filters were discussed in the accessory section, but once again, the most important consideration is *safety*. While the hazards of solar observing have been overstressed in amateur literature, needlessly scaring people out of viewing the Sun, the danger to your eyesight *is* very real. A moment's carelessness can result in your vision being damaged for a lifetime. OK, standard lecture over. If the Sun is treated with respect, it can be a tremendously rewarding subject for observation and study.

What can you see on the Sun? With an inexpensive white-light full aperture solar filter, you'll see, most of all, sunspots: dark spots of speckles on the Sun. Sunspots are far more complex in appearance than you might expect. They are composed of an inner umbra and an outer penumbra, and offer many details to an 8 inch telescope. They can appear singly and in groups, and you can watch them slowly move across the face of the Sun as it rotates (it takes the Sun about one month to complete one revolution at its equator). A white light filter will also, under good conditions, show the granular appearance of the Sun's visible surface, the photosphere. If you're lucky enough to be treated to a partial or total solar eclipse at your location, or can travel to the path of a solar eclipse, you'll be awfully glad to have that full aperture solar filter! A hydrogen alpha filter, as we've discussed previously, is an expensive item for the amateur, but it can reveal even more of the Sun's secrets, including glorious solar prominences along the Sun's limb, and awe-inspiring flares on the disk.

Mercury

Little Mercury, named for the fleet, wing-footed messenger of the gods, is the first stone from the Sun, orbiting at a distance of about 58 million kilometers, making it swiftest of the planets. At a diameter of less than 5,000 kilometers it is also small, the second smallest of the Sun's children (lonely Pluto is now known to be the smallest). Being close to the Sun, it's never far from its master in the sky. As it swings around in orbit, it appears as either a morning star before dawn when it's on one side of its orbit or as an evening star after sunset on the other. Mercury never climbs very high into our skies either. At its greatest *elongation*, its distance from Sol, Mercury is no more than about 30 degrees from our blinding star. With its small size and considerable distance from us, Mercury is understandably very small in telescopes, usually being no more than 5–7 seconds of arc in diameter.

What can an SCT owner see of Mercury? Before you can see anything, you have to *find* this elusive little planet. This is not overly difficult, the main requirements being that you have a good chart, ephemeris or computer program to tell you exactly where it is at any given time, and that you have a clear eastern or western horizon. Trot outside at the proper hour and look to the east or west. Scan the horizon and it won't be long before you find this little world in his appointed spot, three arm's-length fist widths (30 degrees) or less from the edge of the world. You may find a pair of binoculars helpful in locating Mercury the first time, but once you've seen him he'll be unmistakable on further evenings or mornings with no optical aid. Mercury is so easy to see you'll be surprised to learn that if you sight Mercury, you'll be in a fairly exclusive club. The majority of men and women on Earth have never seen (or at least *never recognized*) Mercury due to his early morning/early evening horizon-hugging habits. More surprisingly, there are even quite a few amateur astronomers who've never spotted this distant and rocky world!

Once you know where to look for Mercury, what details can you see with an 8 inch or larger SCT? Sadly, very few. The planet is small, close to the Sun, and always found in the thick and turbulent air near the horizon. Being a *superior* (closer to the Sun than the Earth is) planet, you *will* notice that Mercury goes through phases just like the Earth's Moon, growing from a slim crescent to almost full (the Sun, of course, hides the full Mercury). And that's about it. Mercury's a tiny Moon-shaped thing that moves swiftly into and out of the solar glare.

Can anything help with Mercury? Try to observe as early in the evening and as late before sunrise in the morning as possible to ensure that he's as high up as possible. At these times, the background sky is bright. Eyepiece filters can help a lot here. Red and orange are particularly good for darkening the sky and increasing contrast between Mercury and the background. Images delivered by the Pioneer 10 spacecraft revealed Mercury's surface as a crater covered landscape similar to the Moon. Can we see any hints of these from Earth? It does not appear so. Over the years, visual observers with a variety of telescope types have reported dusky markings on the planet. But these do not appear to correspond to real features as shown in the spacecraft images. But who knows what you might see if you try? For the most part, though, you must be content to just say you've "been" to Mercury, and observe its changing phases as it shuttles back and forth across the sky near the Sun.

Venus

The next planet out from the Sun is Venus. From Aphrodite's beautiful appearance in our morning and evening skies, where she outshines everything except the Moon and Sun, you'd think she'd be a perfect subject for telescopic investigation. This is the Earth's sister, the closet planet in both size and distance to our own world. I remember how excited I was to get my first look at Venus through a telescope. What wonders would be on display? This planet had, up until the end of the 1960s, almost as romantic a reputation as Mars, being imagined as a watery ocean-covered world or a steamy swamp-dominated planet, perhaps inhabited by dinosaurs! What would I see in my 4 inch telescope?

Not much. Through a small instrument, and indeed *any* telescope, Venus turns out to be a severe disappointment. It is really just a larger and brighter version of Mercury. A featureless disk which, due to its status as a superior planet, shows phases like the Moon and Mercury. The featureless nature of Venus, despite its close proximity to the Earth, is due to a deep layer of clouds. The spacecraft that began visiting the planet in the 1960s showed Venus suffers from a bad case of runaway greenhouse effect. The carbon dioxide-laden atmosphere traps heat, resulting in a surface temperature of about 900 degrees Fahrenheit! Sorry, no dinosaurs or Venusian mermaids here.

Is Venus a complete waste of time for the owner of an SCT? No, not completely. It is much less interesting than Jupiter, Saturn or Mars. But there are some things to see. You can, of course, observe the phases, watching Venus grow from a small gibbous disk to a large, thin crescent. You can also try to observe the *ashen light*. If you've ever looked at a beautiful crescent Moon hanging in Earth's skies, you've realized that you can not only see the Sun-illuminated portion of the Moon, but also the night side, which glows feebly, but which is definitely there. This effect is often referred to as "the old Moon in the new Moon's arms." The reason we can see the dark part of the Moon's disk is simple. The bright Earth is in the lunar night skies, illuminating the landscape with reflected sunlight, just as a full Moon is bright enough to light-up the landscape of our own world. Over the years, many reputable observers have noted a similar effect with Venus. In addition to the illuminated part of the disk, the dark portion of the planet can sometimes also be seen glowing faintly. But how is this possible? Venus *has no Moon to light its evenings!*

Nevertheless, it is pretty clear that this faint glowing, this *ashen light*, illuminating the dark hemisphere of Venus is real. In fact, I've seen it myself in recent times (1999) with my 8 inch Schmidt–Cassegrain. I'd imagined I'd seen the effect a time or two, over the years, but never was completely convinced or sure. On this particular evening, though, there was no doubt. The night side of the almost half-illuminated planet was glowing faintly! I was so surprised that I rubbed my eyes and changed eyepieces a couple of times! Couldn't be, could it? Yet there it was. This faint yet obvious glow even remained visible when I added an 80A blue filter to my eyepiece.

What *is* the ashen light? My gut feeling is that though this is a real effect, it is not a real *phenomenon* of the planet. The human eye/mind is a wondrous combination, but is all too prone to allowing us to see what we *expect* to see. Venus looks just like a little Moon, and the brain delivers a little Moon image, complete with earthshine. Combine this "fill in the blanks" characteristic of the eye/brain with effects caused by the high contrast between the brilliantly illuminated planet and the dark sky, and I don't think we have to look too much farther for the reason for the very real ashen light. Still, it is possible, I suppose, that it is an objectively real characteristic of the planet and is caused by high altitude auroral glows or some similar mechanism.

And there are always Venus' elusive markings. By *markings*, you understand, I mean shadings caused by clouds in the planet's impenetrable atmosphere. But don't imagine that these atmospheric features will stand out like the cloud bands of Jupiter. They are *incredibly* faint and subtle, and I have never been completely sure that I've seen *any* real markings. I do occasionally see faint dusky patches along the terminator, but I usually ascribe this to contrast effects. It is possible to *record* details in the atmosphere of the planet, though. A CCD or video camera equipped with an ultraviolet filter and attached to your SCT will definitely show features in Venus' steaming cloud blanket.

How can you maximize your chances of seeing these elusive features? When Venus is overly bright, filters of any color can help reduce the glare. You can also try violet filters in the hopes of bringing out the rare surface detail. If the seeing is steady, you can also use higher magnification to tone down Venus.

How often do I observe this nearby but really *dull* world? A few times an apparition, perhaps. For me, as for most SCT users, Venus is a featureless desert of a world, an object of only occasional interest, rather than a lifelong obsession like the "big three": Mars, Jupiter and Saturn.

Mars

As John Wilford says in his wonderful book, *Mars Beckons!* And not just for the space enthusiasts who dream of the Angry Red Planet as mankind's next home. Mars has fascinated Earthbound observers for centuries. After Jupiter, it is probably the planet that is the most interesting for SCT owners. Unlike Venus, Mars offers detail aplenty: subtle but easily visible surface markings that sometimes change, polar ice caps that grow and shrink, atmospheric clouds, dust storms and more. Plus, there's the fascination of Mars as a *place* to keep us coming back. Though we now know Mars is not exactly the "abode of life" that American astronomer Percival Lowell imagined, it's possible life existed there in the distant past. It's even conceivable that this world, which was at one time much wetter and warmer, still holds some primitive form of life. It's no surprise Mars is one of the first targets to attract the attention of the new CAT owner. Unfortunately, the beginning observer is often tremendously disappointed in the Red Planet.

Mars is terribly fascinating for the visual observer equipped with a small telescope. But it is also just plain difficult. Why? Because it is small and far away. This little world, at about 6,800 kilometers in diameter, barely more than half the size of Earth, orbits the Sun at a distance of about 225 million kilometers. At some points in its orbit, it can be almost 400 million kilometers away. At this distance, it's so tiny that it's barely worth a telescopic glance. Even large instruments will show little or nothing of its mysterious and interesting surface features. If this were all there were to the Mars story, it would probably elicit less interest from amateur astronomers than even bland Venus. But it's *not* the whole story. Every two years, there's a magic *Mars Time.*

Every other year Mars comes to *opposition.* At this time, it is at its closest to Earth and is directly opposite the Sun in the sky, making it well placed for telescopic observation. The distances of Mars' closest approaches depend on exactly where it is in its fairly eccentric elliptical orbit at opposition time. Every 15 years, it comes *really* close, less than 35 million miles. But even a not-so-close opposition is a fine time for Mars observers.

The tiny red speck of a planet grows and grows, reaching a diameter of up to 25 seconds of arc during favorable oppositions. The featureless disk starts showing more and more detail, the legendary dark patches and ice caps popping into view of even little 3 inch MCTs. At opposition, the magnitude of Mars also grows, making the planet positively glaring. During these oppositions it seems as if every telescope on Earth is staring at Mars. And for good reason. There's a tremendous amount to see.

The polar ice caps are easy in a 5 inch or even smaller CAT when the planet is at opposition. Depending upon where the planet is in its orbit, either the north polar ice cap or the southern one will be pointed in our direction, and will be easy to make out as an intensely bright white spot in even the smallest telescopes.

The legendary and romantic dark surface details are what really draws observers, though. There are, as we've long known, no canals. But their lack is more than made up for by the maria of Mars. These subtle dark patches were once thought to represent vegetation, but are now known to be nothing more than areas of the planet that have been at least somewhat scoured clean of dust by Martian winds. These bear watching by amateurs because they can change subtly, and this is of great interest to planetary scientists.

You can even detect atmospheric features on the planet with your 8 inch SCT. In addition to the yellowish clouds that represent dust storms – which can sometimes enshroud the entire planet – there are bluish clouds caused by the planet's weather systems. If your instrument is at least 8 inches in diameter, you also have a fighting chance of seeing Mars' two asteroid-sized moons, Phobos and Deimos. This is far from easy, however, and relatively few Mars watchers, even very experienced ones, have seen these two tiny balls of rock.

How do observers deal with the difficult planet? The two main problems with Mars observing, its small size and its overwhelming glare, are dealt with by using high magnifications. To even *begin* to see much detail on Mars, even at favorable oppositions, requires a power of at least 200×. 300× is even better. These relatively high magnifications increase the image scale, making it easier to detect subtle features. It is a known fact that the human eye is able to make out small details much better at larger image scales, even when these features are theoretically visible with less magnification. Higher power also has the beneficial effect of dimming the planet's brightness. Of course, you'll have to have good atmospheric seeing to be able to utilize the high powers favored by Mars watchers. But don't give up, even on a night when the seeing makes Mars boil and waver. Reduce magnification slightly if you wish, but keep looking. On the most turbulent nights you'll find that things will steady down momentarily, the planet's blurred features will suddenly become crystal clear, and you'll finally realize what Mars observers live for!

There will be times when the seeing is not good enough to support high powers, but the seeing is not bad enough to make Mars observing hopeless. In this case, you can use colored eyepiece filters to reduce the glare. Even when the seeing is steady and you are able to use higher magnifications, you may find particular filters can bring out subtle details unseen in white light. A wratten 21 orange or 25 red can enhance surface markings, while a blue 80A can help bring out the polar caps and atmospheric features. Above all, keep observing the planet. Don't get disappointed and waste valuable oppositions. You'll find by the end of your first Mars apparition you're seeing many things you couldn't begin to detect when you began watching the planet.

Mars riding high in the night sky at opposition is a magnet for the dreams of many. For those of us who dream of traveling there or even of colonizing this little world, our SCTs provide us with a unique opportunity to visit this fabled and secretive world. Sadly, few – if any – of the people now living will have a chance to walk the sands of Mars. It appears that this is to be an honor reserved for our grandchildren or great grandchildren. But with that 8 inch CAT you can at least travel there in spirit and taste a few of the wonders of mankind's second home.

Jupiter

Jupiter is the King of the Planets, and not just because of his enormous girth – this monster planet is 142,984 kilometers in diameter. Jupiter also reigns over the other members of the solar system in the affections of amateur observers because this great ball of gas is just so consistently *interesting*. There's always a tremendous variety of interesting phenomena to see when Jupiter's in the sky. There are the multicolored cloud bands that feature bizarre and amazing spots and *festoons*, huge disturbances in these dark-colored atmospheric upwellings. The

Great Red Spot, the planet's giant storm system, cruises sedately around the planet always drawing the eye of the telescopic observer. Accompanying the planet are Jupiter's four huge Galilean moons (discovered by Galileo when he first turned a telescope to the planet) shuttling back and forth, crossing in front of Jupiter's disk (transits), casting their dark little shadows on his face ("shadow" transits) and disappearing behind his globe (eclipses).

Mars is interesting, but fans of the Red Planet have to wait for every-other-year oppositions before being able to see much. In contrast, Jupiter, although more distant than Mars, orbiting 778,330,000 kilometers from the Sun, is so big that it is a worthy target for CATs anytime it is visible, which is for months at a time every year. Jupiter does vary a little bit in size from a maximum of 50 arc seconds across, but it never gets *much* smaller, making details easy to see even with small telescopes and low magnifications. For example, while the Great Red Spot is currently (2000) a very pale shade of salmon pink and not at all high in contrast, I've been able to easily observe it with a 3.5 inch ETX Maksutov telescope. You can imagine the kind of detail that will be obvious in the average 8 inch SCT!

Indeed, on a good night the amount of detail on Jupiter visible in an 8 or 10 inch CAT can be almost overwhelming. Not only can you see four or five dark cloud belts, but you can see details *in* these belts, ranging from ragged edges to streamers (festoons) impinging into the bright zones which separate the dark bands in the planet's atmosphere. When the Great Red Spot is undergoing one of its darker periods, I've been able to detect not just its oval shape, but tantalizing hints of detail *within* the spot. The four big moons put on a wondrous display in my C8 as well. When the atmosphere reaches maximum steadiness, the four Galileans (the largest, Ganymede, with a diameter of 5,262 km is really planet-sized) are not just star-like points. They show tiny perfect disks! It's easy to spot the hard little shadows these Moons cast on Jupiter's cloud tops as they transit in front of the disk. Under stable seeing conditions it's even possible to track a satellite itself as it transits the planet. The moon will appear as a tiny, bright disk set against the background of mighty Jove.

Many beginners, by the way, wonder why their SCTs show only four moons despite the fact Jupiter is known to have a retinue of at least 16 satellites. The reason is that the 12 remaining moons are all tiny and dim. They are more akin to flying mountains than Galilean worldlets. The brightest of them, Amalthea, the last of Jupiter's companions to be discovered visually (1892), is a dim magnitude 14.1. When this is coupled with the moon's nearness to the glaringly bright disk of the planet, it makes Amalthea a terribly challenging object, even for the largest amateur SCT.

Jupiter is immediately impressive to the new SCT owner. But beyond a pair of bands and the moons, not much will be visible at first. Making out detail on Jupiter is easy compared to the difficulties the other planets present. But it still requires experience and knowing a few tricks of the trade. Foremost – as with all the planets – is a need for precise collimation of your telescope. This makes all the difference in the world. Magnification? Jupiter, being larger, doesn't need as much as Mars in order to give up large amounts of detail. The Great Red Spot and small atmospheric features like spots and festoons are often detectable at powers of 100× by experienced observers. 200× is usually more than enough magnification for this planet, though veteran observers may often push this considerably higher in order to make out the smallest features.

Despite these many wonders, people who've never before seen Jupiter in a telescope are often quite disappointed in their first telescopic glimpse of the King. The planet doesn't look a thing like the Voyager spacecraft pictures! It's all washed out. Where are the dark reds and bright blues and yellows? The truth is that the Voyager images, while amazing, are not a realistic representation of the planet's appearance. Contrast and color saturation in these pictures were strongly boosted in order to make small details stand out. They also make for a photogenic (if somewhat weird-looking, in my opinion) Jupiter. No, Jupiter the real planet, seen live in your eyepiece, is *not* a riot of color. This is a *pastel* planet. Colors are visible to the experienced observer, but they tend to be muted tans, cream-yellows and subtle blues rather than primary colors.

There are things you can do to enhance your Jupiter experience, of course. Filters are a good way to bring out those pastel cloud bands. I find an 80A blue filter a godsend on Jove. It often makes the difference in seeing and not seeing barely visible festoons and whorls. Some observers think eyepiece filters aren't much help with SCTs and other scopes smaller than about 8 inches. But I find they can improve views even in *very* small instruments. I often use filters with my 3 inch refractor. Jupiter is always bright enough to allow filters to work well, even with this tiny telescope.

I've been observing Jupiter for 35 years and he never ceases to amaze me. Just when you think you've seen it all, something dramatic happens. The Great Red Spot fades away, belts disappear and reappear, and long-lived white spots bloom and cruise along the belts. This enormous planet, almost frightening in his majesty, serves to show this is not a static solar system, but one that changes and *lives*. The most wonderful thing is that even a tiny CAT can give you a ring-side seat for Jupiter's ever-changing and never-ending show.

Saturn

Saturn, jewel of the night, is the most beautiful object in the heavens. The first glimpse of this almost artificial-looking world is unforgettable. It's just too perfect to be believed. I've actually had guests at public star parties peer into the corrector end of my scope after viewing Saturn, looking for the photo of the planet they're sure I've "pasted on the end of the telescope!"

Beyond the striking beauty of Saturn's golden-orbed, ringed visage, there's a fair amount of detail available to owners of 8 inch, 5 inch and even smaller CATs. Examine the rings carefully, even with an ETX, and you'll see a thin black line dividing Saturn's A and B (the outer and inner rings, respectively) rings in two. This is the Cassini Division, named for the famous seventeenth-century astronomer who first noted this curious feature. It is caused by gravitational effects that sweep this area clean of ring particles. As most people are aware, the marvelous rings are composed of mountain-to pebble-sized chunks of ice.

The Voyager spacecraft revealed numerous gaps in the rings, all of them much narrower than Cassini's. The only one of these other ring divisions possibly visible from Earth lies almost at the edge of the A ring and is called the Encke or Keeler gap. Earthbound telescopic observers reported a division here over the years, and the Voyagers, and more recently, the Hubble Space Telescope, have imaged it clearly. It is very thin and very hard to detect, however. I've occasionally thought I've caught a glimpse of it in my 8 inch SCT, but what I and other small scope owners are no doubt seeing is the Encke minima, a slight darkening of the A ring near its outer edge rather than the gap itself.

Inward from the B ring is Saturn's final major ring, the C or Crepe ring. The Crepe is often difficult to see in amateur telescopes. It is transparent and appears as a faint dusky haze inside the B ring. Often the easiest way to detect this subtle band is to look for a darkening where the ring passes in front of the planet.

Like Jupiter, Saturn is a gas giant world. A huge ball of gas with (perhaps) a small rocky core at its center. But the appearance of its globe is very different from that of Jupiter. Jupiter is a pastel low-contrast world, but the cloud features on Saturn are even *more* understated. Because of what is apparently a hydrocarbon haze high in the atmosphere, Saturn's belts, spots and zones are much less contrasty than those of Jupiter. Most obvious is a bright equatorial zone. This is flanked on either side of the equator by a slightly tannish region that is fairly easy to detect against the burnished gold of Saturn's globe. This darker area forms, respectively, the north and south equatorial belts. Going higher in latitude in either hemisphere reveals further belts, but the narrow zones which separate them, and the very similar tone of the belts, make it difficult to separate one from another. It often simply appears that the planet above and below the bright equatorial zone is just a uniform color. On particularly steady nights, 8 inch and larger CATs may be able to detect a darkening around the pole of the planet that is pointing toward Earth. This polar "hood" is *very* subtle.

What problems does Saturn present for the SCT owner striving to get a good look at it? The trouble with Saturn is its distance. At 1,429,400,000 kilometers from the Sun, Saturn is small and dim when compared to Jupiter even though its equatorial diameter is a huge 120,536 kilometers. Many new observers, once they get over the initial impact of Saturn's beauty, are distressed by how *small* it is. In order to see details in the ring or on the globe, it really is necessary to kick things up a notch. Where 200× was a good maximum power for hunting Jovian features, this is merely a *starting point* for Saturn. Use as much power on this planet as your scope and seeing conditions will allow. You may try filters to help bring out subtle details – an 80A blue is good – but the dimmer magnitude of Saturn makes filters less usable in small apertures. Luckily, Saturn, for some reason, seems to take magnification much better than Jupiter. On really good nights, I use as much as 600× on the planet with my Ultima C8 SCT. When the air is steady, the planet doesn't break down, it keeps getting bigger and delivering more detail.

Or as much detail as it's capable of, anyway. The fact is once you've examined the planet's wonders thoroughly, there's very little change on Saturn to keep you coming back for more. The less active atmosphere produces fewer things to see on the disk, and the ring is almost unchanging. "Almost" because the rings do change their tilt. The tilt of Saturn to the ecliptic plane, the plane of the Earth's orbit, means that as it and we move in our orbits, the aspect of the rings changes. When they are turned as much as possible toward Earth, they are at an angle of 25 degrees. Eventually, the aspect of the planet changes until the rings are seen *edge on*. Such a *ring plane crossing* took place in 1995. At this time, the rings briefly disappear, allowing earthly observers a good look at the disk and the planet's faint moons without the interfering glow of the ring system. This progression from open rings to closed rings to open again occurs in cycles of 13.7 and 15.2 years. As of this writing, the rings are opening again and are well on the way to maximum tilt as seen from Earth.

At infrequent intervals, which seem to be associated with Saturn's closest approach to the Sun (perihelion) during its 29-year orbit, great storms can erupt on the disk of the planet. These features are usually visible as bright white areas near the equatorial zone and can be seen with even the smallest telescopes. Long-lived white spots can also erupt on the face of the planet, and are monitored by the ALPO observers who keep a watch on the ringed world.

Moons? You want moons? Saturn's got 'em. An amazing retinue of 18 at the last count. Most of them are considerably smaller than Jupiter's four big satellites, but Titan is, at 5,150 km in diameter, planet sized. It even has a thick atmosphere, which is dominated by nitrogen with traces of methane and other gases. Titan is the only moon in the solar system to feature a real atmosphere. The make-up of Titan's air gives it an orangish color easily detectable in an 8 inch SCT. Titan itself is easily visible in a 90 millimeter MCT. In addition to magnitude 8.4 Titan, four other

Saturnian moons, Rhea (magnitude +9), Tethys (10.3), Dione (10.4), and Enceladus (11.8) are easy to see in modest instruments. I've seen Rhea and Tethys in a Meade ETX90. Except when a ring plane crossing is in progress, the positions of the satellites are harder to track than are those of Jupiter's four big moons, since they don't stick to the equatorial plane of the planet as do the Galileans. Computer programs are available to help you identify the positions of the satellites for a given day and time, and I find them much more useful and decipherable than diagrams in magazines.

Saturn just isn't as interesting a world for the CAT-equipped amateur as Jupiter. But it's so beautiful as to be endlessly attractive. Its relative changelessness seems to befit the massive and brooding father of the gods. Even though I know I won't see much different on this distant giant from night to night, I can't help turning my little SCT to Saturn every time it's visible in the night sky.

Uranus, Neptune and Pluto

For SCT users, Uranus, Neptune and Pluto, the solar system's outer trio of planets, are the real "been there" worlds. "Been there" because there's not much of interest with these three, beyond the simple attraction of saying you've tracked down and viewed these objects in the telescope – you've been there.

Uranus

At magnitude 5.8, Uranus isn't much of a challenge for the CAT owner. The planet is actually visible to the naked eye from somewhat dark sites and is quite obvious in a pair of binoculars. The only trick to finding Uranus is knowing exactly where to look. With a tiny disk only 3.6 arc seconds across, this distant gas giant can be easily passed over as just another star. To find the seventh planet, refer to monthly astronomy magazines for its current position in right ascension and declination. With these values in hand, you can plot the location of the planet on your star atlas. Once you're in the right area, you should be able to detect a fairly bright, faintly greenish "star" in the field. It won't show a striking disk at typical finding powers of 100×

or below, but it looks distinctly non-star like even at these magnifications. Once you've centered Uranus, you can run the power up to 200×–300× to get a look at this distant world's tiny face.

Neptune

Neptune is like Uranus for telescopic observers – only more so. The bluish-green globe of the sea god is only slightly smaller than that of brother Uranus, being an immense 49,532 kilometers across, but it is ever farther from father Sun at 4,504 million kilometers out in the darkness! This means that Neptune is both small and dim. Its magnitude is 7.9, so we've left the realm of naked eye objects; you'll need a pair of binoculars or a small telescope to even detect this distant planet as a "star." And if you thought Uranus was small, Neptune makes it look big. The disk of this planet is a tiny 2.9 arc seconds in diameter, meaning that you'll require high powers with your CAT to resolve a disk at all.

As with Uranus, you have to know exactly where to look for Neptune. This isn't very difficult if you're equipped with a good star atlas and a set of coordinates for the planet. The main problem is knowing for sure when you've got Neptune in the field. It does look somewhat non-stellar at modest powers, but it is not as noticeable as Uranus. You may have to examine several candidates at high powers (250× *plus*) before being able to say with certainty that you've visited the solar system's outer gas giant. Don't expect to see much more than an extremely tiny bluish green disk. Neptune does have a much more active atmosphere than Uranus, though, and some visual observers using large, long focal length telescopes have reported seeing details in the atmosphere.

Pluto

Pluto is in a whole other class, both in makeup and in the difficulty you'll have observing it. Unlike the outer gas giants, Pluto is a tiny ball of ice. Once thought to be larger than Mercury, Pluto has been radically down-sized over the years as we've learned more about it. At this time, the accepted diameter of this moon-like world is a mere 2,274 kilometers. It would be tempting to dismiss this little place as an escaped moon of one of the gas giants, but current theories do not support this. And, though it may be moon-like in size and composition (its low density seems to point at ice as its main constituent), Pluto is actually the owner of a moon of its own, Charon, which is just a little smaller than Pluto itself.

Pluto is both small and incredibly distant. On average it is 5,913,520,000 kilometers from the Sun, making it a mere 0.1 arc second in diameter in earthly telescopes. The planet is just not resolvable as a disk by amateur scopes (modern professional telescopes equipped with adaptive optics do have a shot at it). The huge distance makes this small world very dim. At a current magnitude of 13.8 it is completely beyond the reach of many amateur scopes. *Experienced* observers *can* find Pluto with an 8 inch Schmidt–Cassegrain – it is much easier in a 10 or 12 inch SCT, however.

What you'll be looking for is a terribly dim star-like object. It won't be visible in your finder, so you'll have to rely on your main scope to actually pick up Pluto. You'll need several charts to help you track down this little sprite. A wide-field chart will get you in the general neighborhood; a more detailed magnitude 9–10 atlas will help you place the cross-hairs of your finder in the proper spot among the stars. A detailed eyepiece field chart going down to about magnitude 14 (created with a computer star atlas program) will allow you to actually identify the proper field and the planet itself in the eyepiece.

With just a little looking, you should be able to determine whether the stars in the field match what's on your finder chart. If they do, you should be able to identify Pluto. Remember, it will be *very* dim. With an 8 inch telescope it will be quite difficult. Increasing magnification helps bring out dimmer stars by upping contrast between dim object and the sky background. 200x or more may be necessary. You'll also find that the old deep sky observer's trick, averted vision, will help you. To see the very dimmest objects your CAT is capable of showing, don't look directly *at* the target, look *away*, off to one side. A nearby large telescope can also be a tremendous aid. If you can get the owner of a 12 inch or larger instrument to find Pluto, you can compare the views in the two scopes and can track the ice ball down with relative ease in your 8 inch or even smaller telescope. How much aperture does it take to make Pluto "easy?" About 12 inches. I find Pluto "obvious" in my 12, when I'm under dark skies.

How can you be *absolutely* sure that you've seen Pluto, since it looks just like a star? With a detailed finder chart, you can be pretty sure. But the time-honored method of verifying Pluto is to draw a quick sketch of the field. Come back the next evening and check to see if your Pluto candidate has moved with respect to the stars. If so, success!

Why would I devote this much time to helping you locate this admittedly visually uninteresting planet? Because, if you're like me, you'll feel a special thrill in knowing you've found a world that until recent times – until Clyde Tombaugh discovered it in the 1930s – was completely unknown. Tracking down Pluto is also a good test of both your telescope and your locating skills. Most of all, though, is the pleasure in knowing you've gazed upon a world that has been viewed by only a relatively small number of human eyes.

The Rest of the Solar System: Comets and Asteroids

The solar system has sometimes been described as "Jupiter plus debris." While we don't think of Earth or Mars or any of our other little worlds as "left-overs," there is quite a bit of real junk remaining from the formation of the solar system. Comets and asteroids are a further area for the Planetary enthusiast to explore, and an 8 inch SCT can do good work with these little objects.

Comets

Every once in a while a spectacular comet visits the inner solar system. After a 20 year plus drought of brilliant comets, we were recently treated to two "great" ones, Comet Hyakutake and Comet Hale Bopp. The visit of spectacular comets like these two is a particularly exciting and busy time for astronomers both amateur and professional. We're in the spotlight, with the public looking to us for both spectacular views and information. It seems as if suddenly everybody's interested in looking through your CAT. Even your formerly skeptical brother-in-law is no longer puzzled about why you spent "all that money" on your SCT!

Actually, your SCT is not required when a great comet is in full flower; a pair of binoculars will usually suffice. Your SCT can do an incredibly good job when the comet is dimmer, while approaching or moving away. The only bad thing about comets, impressive comets, is that there are so few of them. Another Hale Bopp could be on its way in now, or we could not see its like again for another 20 years, or longer. But, every single year small comets are visible in the skies. Most of these visitors don't get much brighter than magnitude 8, but this makes them perfect for viewing in your SCT. Some are fairly impressive, like the recent Comet Lee, which even showed a hint of a small tail in my C8. Others are mere smudges. But all are interesting. Watch the astronomy magazines and the Internet for news of good comets. Spotting these little fellows can become a nice pastime.

Asteroids

Did you think Pluto was unimpressive? Then you won't like the solar system's minor planets, the asteroids, either. As the name implies, these look just like stars in the eyepiece. The only means of readily identifying them is that they move against the field of background stars. These objects are inherently interesting because of what they are rather than how they look. The asteroids are left over chunks of a planet prevented from forming by the gravitational influence of mighty Jupiter. The area between Mars and Jupiter is littered with these chunks, which range from a few hundred kilometers to a few meters in size. Most interesting for the SCT user is the handful of relatively large and bright minor planets, with Ceres and Vesta being the best of the bunch. Using a finder chart of your own making or one from a magazine, you can hunt down the larger asteroids with relative ease. The prime attraction other than the "been there" factor is just watching their motion against the stars.

Want to discover asteroids? Amateurs are doing it all the time. An SCT equipped with a CCD camera is a perfect tool for this search. Furnished with the proper software to help them determine what's a known asteroid and what isn't, amateur asteroid hunters can use their sensitive electronic CCD cameras to make exposures along the ecliptic. Two pictures of the same field are taken a suitable length of time apart and are compared to see if any objects moved. If one of the "stars" has changed position, the observer knows that likely an asteroid has been captured. The hunt for asteroids is one of the areas where the experienced SCT user can really contribute to science.

Deep Sky Observing

Planetary observing is fun. The bounds of the solar system contain enough space to last some amateurs a lifetime. But there is no denying the call of deep space is heard by many SCT owners. After looking at the Moon and a planet or two, the new CAT owner is eager to see some of the objects that appear in the beautiful color photos in the monthly astronomy magazines: majestic spiral nebulae, great red clouds of gas, blazing globular clusters, and gas-swept nests of newborn stars. Your new SCT can show you all of these marvelous objects, but to avoid disappointment, it's important to understand just what your CAT is capable of dredging out of the darkness.

The 8 or 10 inch Schmidt–Cassegrain is a fine telescope, and it's capable of many things, including delivering satisfying views of deep sky objects. But the pursuit of deep sky objects (DSOs) is, it must be admitted, a game in which aperture almost always wins. A huge 16 inch SCT can't keep up with a 30 inch Dobsonian when it comes to visual deep sky work. But even an 8 inch SCT is capable of showing you *hundreds and hundreds* of distant and lovely objects.

The Deep Sky and Aperture

How do the various sizes of CATs perform on the deep sky? Starting with the smallest, the 90 millimeter to 5 inch telescopes can actually do surprisingly well. They can show you a host of brighter objects, including the entire Messier catalog of galaxies, clusters and nebulae. Most DSOs, and especially galaxies, however, will appear as nothing more than very dim smudges in the eyepieces of these small telescopes.

At 6 inches, things start getting a lot better. A 6 inch telescope will show you stars down to about thirteenth magnitude and will thus turn many globular star clusters from round smudges to balls of stars. 6 inches is a good starting place for the serious visual deep sky observer. The CATs currently available in the 6 inch aperture class are all Maksutov–Cassegrains, and these long focal length CATs are not usually considered "good deep sky scopes." But they can do amazingly well on small objects, bringing their famous high-contrast optics into play. Equipped with focal length reducers, which are becoming available now, a 6 inch MCT could be a better choice than a 5 inch SCT.

An 8 inch SCT, that old standby, improves on the 6 inchers significantly. It goes deeper in magnitude than the 6 inch (because the light gathering power of a telescope is dependent on the area of its mirror or lens, an 8 inch scope delivers nearly twice as much light as a 6 inch). Not only can you see *more objects*, but those you can see show *more details*. Eight inches is a good aperture for the deep sky. It's still portable, but brings enough light gathering power to start making many objects, especially the Messiers, look really good.

A 10 inch increases your deep sky range another considerable step. The number of easily visible galaxies, for example, jumps from hundreds to *thousands*, and, just as when going from 6 to 8 inches, more details are visible in the brighter targets. Unfortunately, the 10 inch size is where SCTs start becoming large and heavy.

It's a shame that 12 and 14 inch SCTs are so huge, because the 12 inch aperture point is where many, many deep space objects stop looking just good and start looking spectacular. The largest currently available SCT, Meade's enormous 16 inch improves things even further.

Many deep sky novices are intensely disappointed when that new (and, they feel, expensive) 8 or 10 inch SCT shows nothing but "faint fuzzies" rather than the detailed and colorful nebulae and galaxies found in photographs in books. The fact is, though, that the eye and the camera are very different mechanisms and deep sky visual observing is quite different from photography. But the eye does have one advantage- – tremendous dynamic range. Often objects will actually look *better* visually than they do in photos.

There is one place where our eyes can't compete with film: color. The human eye's dim light receptors, the rods on our retinas, are quite insensitive to color. The color imaging of our eyes is taken care of by the retina's cones. These receptors register color splendidly, but they don't work well in dim light. This arrangement means most deep sky objects look gray. Particularly difficult for the cones is the dim red light of hydrogen emitted by many nebulae. This is why the beautiful reds and pinks of the Orion Nebula are almost totally invisible in any telescope. Most color films, on the other hand, have no trouble at all picking up this color, even with rather short exposures. This is quite disappointing for new observers, but the fact is, the brightest deep sky objects are usually too dim for our cones, and look gray.

A few, a very few, of the very brightest DSOs can occasionally reveal a *little* color to the experienced observer using an amateur-sized CAT. I've seen the Orion Nebula appear as a striking green in an 8 inch SCT when conditions were exceptional, and I've seen dim brownish-red tinges around the edges of this same object with a C14. Some small, bright planetary nebulae also have enough "oomph" to stimulate the eye's color sensitive receptors. The Saturn Nebula in Aquarius is easily seen as an electric blue color in rather small apertures. Please realize, too, that some colors you see in photographs have been computer manipulated in order to have more impact; they would not be so bright and gaudy in a normal photo.

Finding Deep Sky Objects

An 8 inch SCT, even one stationed in your light-polluted suburban backyard can show you dozens and dozens of DSOs that will literally take your breath away. *But only if you can find them*, and therein lies the biggest problem for the new deep sky observer. How can you find something you can't see? Many deep sky objects will be invisible or practically invisible in your finder scope, especially in light-polluted skies. There are two principal methods of finding dim objects in the sky. The first method, star hopping, uses your eyes, your finder scope and your star charts to track down objects. Then there's the automated method, using setting circles, either the analog setting circles that come with your telescope or add-on digital setting circles (DSCs), to locate targets.

Star Hopping

Star hopping is a very low-tech operation, requiring only your two eyes, a good star atlas, knowledge of the constellations (which I hope you accumulated before you bought your first telescope) and a little knowledge of directions and distances in the sky. Your telescope came with at least a 30 millimeter finder, which can work well for star hopping, and we'll assume you've taken the advice offered in the Accessories section of this book and purchased a detailed star atlas. So all that remains before we start hopping our way to deep sky bliss is a little talk about directions and distances.

Directions in the sky can seem confusing at first. But, finding north, south, east and west in your telescope field is really easy. In any telescope, east is the direction in the field of view where stars enter the field when the clock drive is turned off, or where they enter when you move the scope toward the eastern horizon with your slow-motion control. West is where the stars exit the field of view of finder or main scope when you do this or when you shut off the drive. North? It's the direction in the field of view where stars enter when you use the declination slow-motion control to nudge the scope toward the northern horizon. The direction where stars exit the field is south. That's all there is to it. You don't have to worry about the difference between an inverting finder and your star diagonal equipped main scope if you just keep this simple layout in mind.

Many beginners also find the concept of relative positions of objects hard to master. It's really easy to establish where a DSO or star lies relative to another object, though. An object is west of another object if it's farther toward the western horizon. It will also rise before an object to its east. One star is north of another if it is closer to the northern horizon than the second object; the reverse is true if it is south of the second object. If a star is farther toward the northern horizon and also toward the western horizon than another, it is northwest of the first star. To keep from becoming confused, just keep in mind that all we're talking about is *which horizon(s) one object is closer to than another.*

Distances in the sky are even easier to determine than directions. The sky is divided into 360 degrees of arc. One degree is further divided into 60 minutes, and 1 minute of arc is made up of 60 seconds. Beginners are often confused when they find a galaxy described as being 45" across. 45 inches across? No, 45 seconds. A double quote mark (") is used to denote seconds of arc, a single quote (') is minutes, and the usual symbol is used to represent degrees. You can make rough judgments of distances in the sky using your hand. Your outspread hand covers about 15 degrees of sky when held at arm's length. A clenched fist, as we mentioned earlier, covers 10 degrees. Your index finger is about ½ degree in width.

It is often useful to know how much area of the sky your finder's field covers. Find two stars which just fit across the diameter of its field. Then, measure these stars with your fist or fingers to find out how far apart they are in degrees of arc. If you want to be a little more accurate, you can reference these stars in your atlas. Measuring the distance between the two stars against the declination scale (whether the stars are north/south of each other or not) will tell you precisely how far apart they are in the sky, and, thus, how much field your finder telescope covers. How about the field of an eyepiece? The true field of sky that the eyepiece covers?

The easiest way to determine the distance scale or true field of an eyepiece is with this formula: true field = apparent field/magnification. Let's say, for example, that you have a 25 millimeter eyepiece and you want to find out how big its true field of view is. We know that its magnification with our f/10 SCT is 80 (2,000 millimeter focal length divided by 25 millimeters). We then reference the manufacturer's specifications of our eyepiece to determine what its apparent field is. We find that our Kellner is listed as having a 50 degree apparent field. Doing the math, 50 divided by 80, we come out with 0.62 degree, or 37 minutes of arc. Just judging by eye will allow you to determine smaller distances in your eyepiece. The distance from the edge to the center is a bit less than 20 minutes of arc, half this is around 10 minutes of arc, and so on.

Equipped with this important knowledge of directions and distances, you'll be able to hop to any object you desire. Let's use the beautiful globular star cluster M15 as a test run. Page through your star atlas until you find the huge constellation Pegasus, the Flying Horse. It's easy enough to recognize both in an atlas and in the sky by the giant Great Square formed by four prominent stars. M15 lies in the western part of the constellation not far from the bright star Enif (Epsilon Pegasi), about 20 degrees west of the Square. This star, which forms the "horse's nose," will be our guidepost to M15, which lies at an RA of 21 h 29 m and a declination of +12 degrees 10 minutes.

Looking at your chart again, you'll see that a fairly prominent star, magnitude 3.7 Theta Pegasi, lies about 7.5 degrees to the southeast of Enif. Draw an imaginary line through Theta, through Enif, and keep going for about two-thirds of the distance again between Enif and Theta (about 4 degrees) along a straight line and you'll run smack into M15 (see Figure 7.2).

Study this pair of stars and the marked location of M15 on your chart before going to the telescope until you've got the position of M15 well-fixed in your mind. At your CAT, insert your lowest power eyepiece and loosen your RA and declination locks until the scope can move freely, but is not so loose that the tube moves suddenly if you let it go. As always, grasp the tube near the corrector end and near the rear cell, and swing the OTA in the direction of Theta. When you get in the general area, put your eye to your finder scope and gently guide the CAT until the cross-hairs lie on Theta Pegasi. Have trouble getting on Theta? Use the trick of keeping the eye not at the finder eyepiece open. With a little practice you can learn to "superimpose" the telescopic and nontelescopic images, making it easy to get on the star. Referring to your chart (remember, you must usually invert the star atlas so that it matches the

Figure 7.2. Bright stars and groups of stars provide sign-posts for hopping to dim deep sky objects.

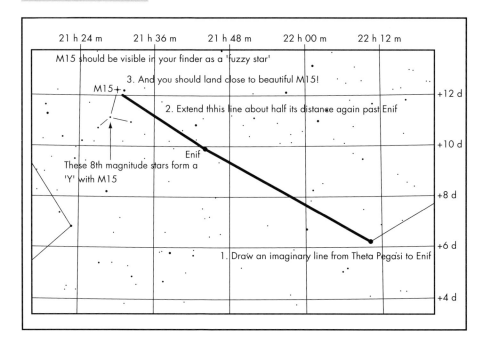

- M15 should be visible in your finder as a 'fuzzy star'
- 3. And you should land close to beautiful M15!
- M15+•
- 2. Extend thhis line about half its distance again past Enif
- Enif
- These 8th magnitude stars form a 'Y' with M15
- 1. Draw an imaginary line from Theta Pegasi to Enif

21 h 24 m   21 h 36 m   21 h 48 m   22 h 00 m   22 h 12 m

+12 d
+10 d
+8 d
+6 d
+4 d

upside down view delivered by most finders). Move the scope slowly and gently to the northwest until you get to bright Enif. Keep going, taking note of how far you traveled from Theta, until you've covered another four degrees of distance. In this area you'll find a semi-prominent (in your finder scope) magnitude 6 star. This little star, which is just outside the line drawn from Enif, is right next to M15.

Lock the telescope securely in both axes, taking care not to move it. Put your eye to the eyepiece. Somewhere in the field, you should see a small, bright and obviously non-stellar glow of little M15, a magnitude 6.3 globular cluster. If you don't, slew back and forth in declination and right ascension using the slow-motion controls (remember to loosen the RA lock before using the slow-motion control) until you have the cluster in the field. You may then center it and insert a higher-magnification eyepiece to help resolve some of the cluster's many tiny stars (in your 8 inch or larger telescope).

What if nothing's there, and hunting around with the slow-motion controls doesn't bring M15 into view? Resist the urge to swing the scope wildly around in RA and declination – this usually won't help. Instead leave the CAT where it is and check your finder's view against your chart. Did you go far enough from Enif? About two-thirds the distance between Theta and Enif past Enif? Or maybe too far? Did you remember to stay on the imaginary line between Enif and Theta? Did you properly identify Enif and Theta when you started? Glance up at the prominent Flying Horse and satisfy yourself that you're in the right part of the constellation. Usually, when I can't find an object after repeated attempts, I find that I'm not just a little off – I'm usually *many degrees* from the proper area of the sky.

If M15 still fails to put in an appearance, another possibility is that you're using too low power an eyepiece to make this star cluster show up well. In light-polluted areas, the sky will not be black, but a shade of gray. Even from dark skies, too low a power may make an inherently small object like M15 (12 minutes of arc across) appear star-like – you may have accidentally passed it over, thinking it was just another star. Go to a higher-power eyepiece and start over again. Search carefully and you will be rewarded. Just take things slowly and remember, few deep sky objects will be bright enough, especially from suburban skies, to jump out at you. Examine each field minutely before searching farther afield. Once you've located M15 (and you will), take a deep breath, relax and just take a good *long* look at this wondrous ball of thousands of ancient stars. This look at an incredibly old star city is the real reward for your searching and for all those weeks and months when you wondered about a telescope, searched for a telescope and waited for a telescope to be delivered!

That's all there is to star hopping. Use your star atlas to find prominent stars and groups of stars to lead you to the target. You can start out planning your hops in detail indoors, but you'll soon find that routing star hops is something you can do on the fly in the field with atlas and scope.

Analog Setting Circles

Is star hopping the only way to find objects? No, you can use (or at least try to use) your SCT's setting circles to automate object location. The key to making effective use of setting circles is *polar alignment*. For your setting circles to lead you to the objects you "dial in," you must have your telescope precisely polar aligned. Closer to the pole, that is, than merely pointing your telescope's right ascension axis at the North Star. This is not a big hindrance to using setting circles, especially if you have a polar alignment scope or other pole finding aid, or can do a drift alignment. See Chapter 10 for instructions on performing a very accurate polar alignment.

Before using your setting circles, you must *calibrate* the RA circle (see Figure 7.3). This is easy to do. Choose a bright star whose right ascension coordinates you know (the owners' manuals for most SCTs include a list of bright stars and their coordinates) and put it in the center of a high-power eyepiece (your clock drive should be running). Now, set the RA circle on your scope to the RA of this star. Fork-mount SCT RA circles are free to move, so you just slide the circle along until it is reading the correct value in hours and minutes under the pointer on your scope's base. For example, if you've chosen Sirius as your alignment star, you'll move the RA circle until it reads 6 h 45 m, about three-quarters of the way between 6 h and 7 h on your circle.

Figure 7.3. The right ascension setting circle must be calibrated before each use.

You may be surprised and confused to find that your right ascension circle has two different scales, one inner scale and one outer. Which to use? One scale, usually the inner scale, is for use in the northern hemisphere and the other is used south of the equator. Your manual should identify which is which, but if it doesn't, use the scale where RA increases under the pointer as you move the CAT toward the eastern horizon. If your scope is equipped with a German equatorial mount, you may have to loosen a set-screw to set your RA circle, but otherwise, the process is the same. Once you've got your circle set, it will remain correct with most telescope mounts *unless you turn off the clock drive.* If you do shut off the drive for some reason, you'll have to realign the circle on a bright star. The declination circle is set at the factory and should rarely require user alignment.

Let's make delicious M15 an example again. Set the telescope in RA first. Unlock your RA lock and, using either the slow-motion control or by just moving the fork by hand, rotate the telescope in RA until the pointer indicates a value of 21 h 29 m (just a hair under 30). Lock the axis securely when you've got the scope set in right ascension. Undo the declination lock part way and move the telescope by hand. You'll notice that your declination circle goes from 0 to 90 and then repeats. To get on the proper scale, move the scope in declination until it's obviously pointing south or north, depending on the declination of your target (southerly declinations are usually expressed with a minus sign as in −30), and slowly move it to the proper value. On just about all fork-mounted telescopes, the declination circle is significantly smaller than the RA and the scale is marked in one-degree increments making it a bit hard to adjust the scope finely in declination. Do the best you can in setting the mount to M15's declination, 12 degrees 10 minutes (just a tiny bit past 12 degrees). When done, take a look through the eyepiece.

Is M15 there? It should be, in a *low-power wide-angle eyepiece*, if you've accurately calibrated your RA circle and if you've been as precise as possible about setting in the coordinates. If it's not in view, don't panic. The first thing to do is to slew back and forth in *declination*. Due to the smaller size of your declination circle, any error is likely to be here. If this doesn't make the cluster appear, set declination back to the proper value and slew about a bit in RA. If you *still* don't see M15, step back for a moment and take a look at the general direction in the sky where your SCT's pointing. Does it seem to be facing in the right direction? If so, check your coordinates again to make sure you entered them correctly and that you have the proper RA and declination for the object of your desire. If this seems in order, you may want to try realigning the RA circle on a star closer to the target object than your original calibration star.

Digital Setting Circles

In the opinions of many amateurs, analog setting circles are old fashioned. DSCs, digital setting circles, are all the rage among CAT users these days. A DSC unit consists of a small computer box and two optical encoders (see Figure 7.4). One encoder mounts on the declination axis and one on the RA shaft. Moving the scope in either axis causes the shaft of the encoder to turn, sending a series of pulses to the computer, which sorts out these pulses into position and direction of movement. DSCs can be very easy to use. Most units don't even require an accurate polar alignment. The majority of DSC computers now also have at least a small selection of objects in memory (typically, the Messier catalog). With these computers, you use the front panel buttons to enter your object's catalog number. The DSC then indicates the directions you should move the telescope in RA.

How do you set up DSCs? And how well do they work? The first thing you have to do is install the RA and declination encoders, the first time out, anyway. Even DSC units shipped with new telescopes typically have to be installed by the user. This is usually an easy procedure, requiring at most a screwdriver and a pair of pliers. Some DSCs for fork-mounted telescopes do require that you remove the RA encoder if you want to store the telescope with the corrector end of the tube pointed at the scope base. In these installations, there's not enough clearance to do this with the right ascension encoder in place – it sticks up from the base too far. But this usually involves merely loosening a set-screw on the encoder shaft.

After the encoders are correctly installed, most SCT users polar align the mounting as well as they can. While this is not *required* for most DSC computers, it can improve accuracy. The next step is to align the *computer*, which is usually done by pointing the telescope at two stars from a list in computer memory. As when aligning a goto instrument, it is important to follow manufacturer's guidelines as to how far apart these stars should be in the sky for the best alignment. Once the second alignment star is entered, usually just by pushing a button, the user will be presented with a "warp factor," a number representing the accuracy of the alignment. If this figure is within usable tolerance stated in the DSC instructions, you can begin using your computer (usually a smaller number is best). If not, you may have to redo your alignment, possibly using different stars.

You should now be able to find any object within range of your scope with ease. You can either use the built in DSC catalog, or just use *coordinate mode* where your DSCs read out the RA and declination coordinates of the scope's current position. You move the scope appropriately until the value of the object you're seeking appears. Either way you use them, digital setting circles are surprisingly accurate when used on SCTs (they may have problems on homemade wooden Dobsonian mounts). Once prop-

Figure 7.4. Digital setting circle computers are all the rage among CAT users these days (photo courtesy of Meade Instruments Corporation).

erly installed and aligned, most computers will put an object in the field of a low/medium-power eyepiece every time.

The Deep Sky

You've decided on an approach to finding objects. But what do you find? Which objects are worthy of your attention? The sky is filled with beautiful deep sky wonders, but the Messier list is the time-honored spot for beginners to start on the deep sky. These 110 objects, discovered by Charles Messier and others, are the best and the brightest. Once you've done the Messier list, you'll probably want to move on to the huge (approximately 8,000 objects) New General Catalog (NGC) list. This list of objects, published by John Dreyer in 1888, was partly based on work done by Sir William Herschel. The objects in the NGC range from Messier class in brightness and detail to objects that are a visual challenge for even the largest CATs. What can you expect of the various types of deep sky objects in either catalog? What kinds of deep sky objects *are there?*

Galaxies

Galaxies, massive island universes, the sisters of our own Milky Way, are the object of many an amateur astronomer's desires. Beginners, particularly, long to see the beautiful pinwheel-like spiral arms that many galaxies display. But there's a problem: galaxies lie far outside our Milky Way and are the most distant objects in the cosmic zoo. The nearest large galaxy, M31 in Andromeda, is a staggering 2.8 million light-years away. These huge distances mean that most galaxies are small and dim. They are also badly affected by light pollution. As we know from our earlier discussion of light pollution reduction filters, they are not helped *at all* by LPR filters, either. Oh, you can see many, many galaxies from urban and suburban sites, but to have a *prayer* of seeing spiral arms with your SCT you simply must get to a dark site. A trip to the boondocks with your CAT can be very rewarding, though. There under the velvety black skies, you can get a true idea of the scope and majesty of your universe.

The Best of the Best Galaxies for Visual Observers

The following galaxies are but a small sampling of what's available to the owner of a 5 or 8 inch CAT, but in my opinion they are what galaxy hunting is all about!

M51 is known as the Whirlpool Galaxy. If you want to see spiral arms visually, this is where you go. This is an interesting and fairly bright galaxy located near the Big Dipper (or Plough) in the constellation Canes Venatici. Despite its face-on orientation, this spiral is prominent enough at magnitude 8.7 to be dramatically visible in CATs. I've easily detected this object's spiral structure from dark sites with a 5 inch telescope. Also interesting is the little irregular galaxy NGC 5195 which lies just to the north of M51. A bridge of stars seems to connect the two galaxies and would seem to be evidence of a recent interaction. This "pulled-off" stream of stars between the two is visible without too much difficulty in 10–12 inch scopes. As is typical for all galaxies, these details are obvious from *dark skies only*. From the typical suburban neighborhood, all that's visible of these two wonders is a pair of dim blobs.

M31 is the most easily found island universe, being visible to the naked eye from even semi-dark sites among the stars of Andromeda. Beginners usually expect a lot from Andromeda, as M31 is usually called, since it is so bright (magnitude 3.5). But they are also usually very disappointed by this galaxy's appearance in a telescope. It appears as nothing more than a bright elongated blob. Why? One difficulty is M31's sheer size. At about 3 degrees across, it's impossible to fit this monster in one field of view using even very long focal length eyepieces with an f/10 telescope. Another problem with Andromeda is her inclination. This galaxy is tilted only 6 degrees from our line of sight, meaning only hints of the arms are visible. Nevertheless, M31 can be an amazing object for the experienced observer, showing off a couple of dark lanes, a giant cluster of stars and a pair of small satellite galaxies, the little ellipticals M32 and M110.

M87 is a monstrously huge elliptical galaxy with a mass of trillions of suns – it is far larger than our own galactic home, having grown so fat, perhaps, by devouring its fellow galaxies in the dense Virgo cluster where it resides. Like most ellipticals, which are essentially huge round balls of stars, M87 isn't much to look at visually, admittedly. But it is the best of the ellipticals. And it is interesting at least for what it represents, a really gigantic star city. CAT users with telescopes in the 12 to 16 inch range may have some hope of seeing a hint of the jet which extends from this strange galaxy.

M104, a magnitude 9.0 spiral, is probably our best example of an edge-on oriented galaxy. Even 5 inch scopes stationed under dark skies can reveal not only its thin sliver of light, but the big central bulge which gives this object its name, the Sombrero Galaxy. In addition to these features, an 8 inch telescope also shows that M104 is bisected lengthwise by a dark lane of dust. In photographs and in really big CATs, this lane is shown to have irregular edges.

Nebulae

Nebulae are the huge clouds of dust and gas that lurk in interstellar space. Like galaxies, nebulae can be very different from each other. Bright nebulae can be divided into four different and distinct types. Emission nebulae, reflection nebulae, planetary nebulae and supernova remnants.

An emission or diffuse nebula is a giant cloud of (mainly) hydrogen gas. These are the great stretches of hydrogen in our Galaxy that give birth to new generations of stars when they contract, due to gravitational effects and the shockwaves of nearby supernovae (these exploding stars also contribute the heavier-than-hydrogen elements that go into our current generation of stars). Until stars are born, diffuse nebulae are dark objects – there are no stars to illuminate them. When hot and massive young stars begin to fuse in the midst of diffuse nebulae, the gas becomes excited by ultraviolet light and begins to glow. Diffuse nebulae are among the most beautiful objects in the heavens.

Reflection nebulae are also a type of diffuse nebulae, but they do not emit light. They are composed mainly of dust, rather than gas and only shine by reflecting the light of nearby stars. For this reason, these objects appear blue in photographs. Emission nebulae glow with the red light of hydrogen.

Planetary nebulae are entirely different from diffuse nebulae. A planetary nebula is the *corpse of a star*. A star in the size range of our own Sun does not explode violently as a supernova, it instead undergoes a lingering death, inflating to red giant size, and eventually ending the thermonuclear fusion that has been taking place since its birth. When fusion stops, the star is truly dead. What's left is a cinder, the remnants of the star's core. This white dwarf forms our planetary's central star. The nebula part of the planetary is composed of the remains of the outer layers of the star from its red giant phase. With fusion done, these layers slowly expand and dissipate in interstellar space. These layers, which form a spherical or tubular shaped shell around the white dwarf, give planetary nebulae a roundish shape that, to William Herschel resembled a planet (Herschel knew these were not planets).

A supernova remnant is composed of the remains of a huge star that has died a violent death. An expanding cloud of gas with a tiny and dim neutron star or pulsar at its heart is all that is left of a once glorious sun. Supernova remnants tend to be fairly dim.

The Best Nebulae

Diffuse nebulae can be affected as severely by light pollution as galaxies. The biggest examples can be especially hard to see from suburban and urban sites. But the amateur has a wonderful tool to help with these objects: the light pollution filter. If you're interested in nebulae, you owe it to yourself to invest in one of these devices, as they can make the difference between seeing and not seeing these clouds for observers equipped with CAT-sized telescopes. In contrast, the small size of many planetary nebulae makes them bright and easy to pick out from even poor skies. They can still benefit from a good filter, however, when it comes to teasing out details.

M42, that great glowing mass in Orion's sword, is the most wonderful nebula in the skies – for northern hemisphere observers. Some would say it's the most beautiful deep sky object of all. It's easily visible to the naked eye, and its size of 1 degree across means that it's not too large to be appreciated in long focal length SCTs and MCTs. In fact, the brightest areas are perfectly framed in a 0.5 degree field of view. It is flanked by a small detached comma-shaped patch of nebulosity which has its own M-number, M43. The Great Orion Nebula looks beautiful in *all* telescopes from the largest to the smallest, and deserves extended study of not only the gas cloud but of the fascinating stars embedded within it. Of particular note is the Trapezium, a multiple star system which reveals four prominent stars to even 90 millimeter CATs and six members to 8 inchers.

NGC 2070: south-of-the-equator astronomers, in addition to getting a really good look at M42 (its southern declination places it nice and high in the sky for southern observers), have another great nebula to marvel over, the Tarantula Nebula located in the far southern constellation of Dorado. This huge glowing cloud stretches 40 minutes of arc across the sky. It is, in fact, a much larger object that is farther away than M42. If it were located at the same distance as Orion, it would cover nearly 30 degrees of sky!

M78, located around a pair of dim stars not far from Orion's belt, is the best example of a reflection nebula in the sky. It is tiny at 8 arc minutes across, but this makes it show up easily, even given its rather dim magnitude of 11.0. Don't expect to see the blues of photographed reflection nebulosity, though. All you'll see is a dim gray cloud surrounding two unprepossessing stars.

M57, the famous Ring Nebula, is but one of a multitude of beautiful planetaries which litter our skies. This dead star is located in the small but prominent constellation Lyra, the Lyre, and is bright and unmistakable at magnitude 8.7 and a bit larger than 1 minute of arc across. A 90 millimeter MCT will reveal M57 without effort as a tiny spot of light. High magnifications and good seeing conditions can reveal at least hints of the doughnut middle of M57 in an ETX or a Questar, but it generally takes a C5 to show the ring shape clearly. An 8 inch SCT will show the ring *very* clearly and will reveal it is not round but somewhat elongated. An 8 inch will also make clear that the middle of the ring is not dark, but a gray color. SCTs in the 12 inch class may reveal the Ring's central star, a magnitude 15 white dwarf. But this is tricky; not only is this star dim, but it is possibly variable.

M1, the Crab Nebula, the first object in Charles Messier's catalog, is the best and brightest example of a supernova remnant in the sky. This object, which appears as a small ninth magnitude oval glow, is easily found in the prominent zodiacal constellation Taurus. Large SCTs may show hints of the strange tendrils of gas that give this nebula its name. This expanding cloud of gaseous debris is the result of a supernova that exploded in 1054.

Globular Clusters

Globular star clusters are incredibly ancient balls of stars. They are thousands of light-years across, and contain from thousands to millions of very old suns. They orbit the nucleus of our Galaxy and are so ancient that they were possibly witness to the birth of the Milky Way itself. Globulars are one of the best reasons for buying a larger-aperture telescope. Even the brightest globulars are composed mostly of magnitude 13 and dimmer stars, so at least a 6 inch telescope is needed to revolve the best clusters, and considerably larger mirrors are required to resolve many of these objects into balls of stars. Our old reliable 8 inch SCT is a decent globular hunter; under dark skies it can provide good resolution on many clusters.

M13, the Great Cluster in Hercules, is the most beautiful globular of them all for northern hemisphere observers. At magnitude 5.8 and 20 minutes of arc in diameter, this object is incredibly prominent. But, while bright, it is not necessarily the easiest globular to resolve in small telescopes. The stars are fairly tightly packed and can be difficult to separate in small telescopes.

Omega Centauri, aka NGC 5139, is *the greatest*. M13 *pales* beside it. Its statistics tell you why. With a magnitude of 3.9, it's easy to see why it received the Bayer Letter Omega, an identifier usually reserved for stars – it is clearly visible from modestly dark sites as a dim-seeming "star." And it is big. At 53 minutes of arc across, this indescribable beauty looks better in finder scopes than many globular clusters do in the main telescope. Resolution? Picking out stars is a snap, even in the smallest CATs. The sad thing for northern hemisphere observers is that this really is a southern object. At –47 degrees south, it is completely invisible from many parts of the far northern US and Europe. At my latitude of 31 degrees north, I can get a nice view of Omega, but I can only imagine what it would look like if it were high in the sky! Ah, well, some day.

Open Clusters

An open cluster is a nursery of young stars. Stars are born in clouds of gas, and once this dissipates, what is left behind is a group of young sparklers. Their movements will eventually cause the cluster to disperse, but for a time they present us with lovely groupings. Open clusters, which are also known as *galactic* star clusters, are as different from globular clusters as can be. Open clusters are composed of the very youngest stars; often they are no more than a few million years old. Globular stars count their ages in *billions* of years. Open clusters are essentially formless groupings, rather than the well-defined balls of stars the globulars arrange themselves into. And the galactic clusters are made incredibly lustrous by the presence of young, hot and massive blue and white stars, those of spectral types O and B. In the globulars, these short-lived stars passed away long, long ago.

Most observers would rate galactic clusters among the least interesting of the DSOs. After all, they appear as nothing more than fairly loose groupings of stars. In the denser portions of the Milky Way, it's often hard to pick open clusters out from the general stellar background. But there are some beautiful examples. And, like planetary nebulae, open clusters have the advantage of being relatively unaffected by light pollution.

M37, located in the winter constellation Auriga the Charioteer, is my personal favorite. At magnitude 5.6 and 21 minutes of arc across it is both bright and reasonably compact. It is also incredibly rich. A telescope that can reach down to around magnitude 12 will reveal at least 150 stars. This cluster is made even more beautiful by the presence of a lone reddish-orange star near the heart of the group. Set off by the cluster's mostly blue and white suns, this provides a wonderful contrast and an unbelievably wonderful experience for the visual observer.

M45, the Pleiades, are, like M31, hurt by size. This cluster is almost 2 degrees across and you may feel that your finder delivers a better view of it than your f/10 SCT. Nevertheless, you can still get a nice image of the rich field of this naked-eye group at your scope's lowest magnification (the brightest of the "Seven Sisters" that make up the naked-eye component of the cluster are at around magnitude 3). Your CAT will naturally reveal many more stars here, and your lowest-power widest-field eyepiece can provide a pleasing view. On a really good night, your main scope may even allow you a glimpse of the tenuous reflection nebulae that surrounds the star Merope and several of the cluster's other bright suns.

M46, in the southern constellation of Puppis, is a nice open cluster available to both northern and southern hemisphere observers. At magnitude 6 and 27 minutes of arc in size, the cluster is fairly rich and interesting in itself. But there's a special treat here for the owners of 5 inch and larger CATs: the tiny planetary nebula NGC 2438. At just over 1 minute of arc in diameter, it needs high power to make it pop out from the cluster stars, but at magnitude 10 it takes magnification quite well. Two nice deep sky objects for the price of one!

Whatever type of observing you decide to do, from solar viewing to galaxy chasing, the important thing is to *just keep going. Don't let the incredible excitement of this first night out with your SCT die away.* As you finish this first evening's observing run, removing your beloved scope from its wedge or mount just as you did when you first set it up for inspection indoors, you should feel tired but elated. You may even be a bit overwhelmed by what you have seen and the certain knowledge that you've only scratched the surface of the solar system or the deep sky. And you should resolve to get outside with your wonderful CAT *every single clear night* if possible. Only by observing frequently can you really learn to "see" and become adept at using the wonderful new instrument which will become your beloved "magic starship" for years and years of voyaging.

Chapter 8

Care and Feeding of a CAT: Alignment, Testing, Maintenance, and Troubleshooting

You can't expect an automobile to go hundreds of thousands of miles without a tune-up, and you can't expect your CAT to put in night after night under the stars without some maintenance. The SCT is remarkably trouble-free when compared to other telescope types, but it still needs occasional adjustments. Making sure your telescope stays in good shape requires learning how to identify and deal with problems of an optical, mechanical and electrical nature.

Collimation

Collimation, the procedure for adjusting the mirrors (or lenses) in a telescope to ensure their proper alignment, is a scary prospect for many beginners. Experienced amateurs tell novices they must collimate their scopes if they're to perform well. But this process seems extremely difficult and, if anything, a good way to ruin a new SCT. Not true. Oh, collimation *is* vital, but there's no reason to fear it. Collimating an SCT is very easy once you get the hang of it.

Because of the magnifying characteristics of an SCT secondary mirror, even a slight misalignment of the primary and secondary mirrors can absolutely wreck images. This is especially true for high-power planetary work. Sadly, many SCT users never check the collimation of their telescopes, don't do it the right way or don't do it very frequently. In my opinion, this is what has given the Schmidt–Cassegrain a reputation as a poor planetary telescope. But a collimated Schmidt–Cassegrain can perform *incredibly well* on the planets.

Does your SCT (or MCT) need collimation? It's true that most CATs, especially those of 8 inches and smaller apertures, hold their optical alignment very well, but you should assume that the SCT's collimation *will* have shifted during shipment if you bought by mail. A road trip to a dark site over rough roads can also throw off secondary alignment. Even if you keep your CAT at home, it's wise to check collimation every once in a while. As for me, I check collimation every time I prepare for a critical observing run – high-power planetary observing, for example.

There are a number of methods for checking optical alignment, but the easiest and best way is with the star test. This is a good thing. Unlike the Newtonian owner, the SCT fan doesn't have to buy any collimation tools. All you need is your eye, a medium/high-powered eyepiece and a bright star. Magnitude 2 Polaris makes a perfect star for collimation checks – I find a really bright star like Sirius difficult to work with.

SCT collimation is a three-step process. You perform a rough collimation check to see if the secondary mirror is at least close to proper alignment. You adjust the secondary if necessary, and then move on to fine-tuning, where you tweak the mirror until your secondary is exactly aligned. If conditions permit, you can make a final check and even more precise adjustment via an in-focus star test.

Rough Collimation

Set up the SCT as usual, insert a medium-power eyepiece, one that will yield a magnification of around 160×, and point the telescope at Polaris. It is the correct brightness and will remain in the field of view without the need for you to make adjustments, even if your polar alignment is somewhat off. When you have it in the center of your eyepiece field, defocus Polaris, turning the focus control either way you choose, until

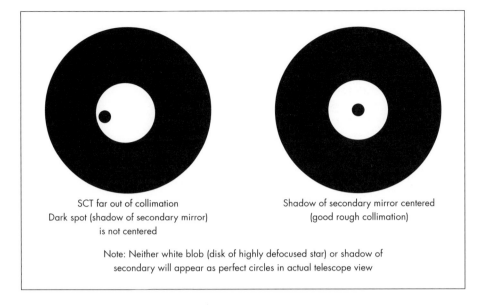

SCT far out of collimation
Dark spot (shadow of secondary mirror)
is not centered

Shadow of secondary mirror centered
(good rough collimation)

Note: Neither white blob (disk of highly defocused star) or shadow of
secondary will appear as perfect circles in actual telescope view

Figure 8.1. To do rough collimation, defocus a star until it looks like a big, round globe of light.

the star expands into what looks like a round globe or blob of light with a dark center (see Figure 8.1). Keep moving your focus control until the star covers about a quarter of your field. Does the dark spot, the shadow of your secondary mirror, seem more or less centered? If it does, move on to the next step. If it does not, you'll need to adjust the secondary.

Some older Celestron telescopes have an orange plastic cover on the secondary mounting. It must be removed to expose the three screws that allow you to adjust the mirror. This cover is held in place by two plastic tabs inserted into the secondary mounting assembly and is removed by gently popping it off. If you've never had this cover off, you may need to gently pry at the tabs with a screwdriver until it comes free. Just remember the cardinal rule of telescope maintenance: *never force anything!* With the cover – if your secondary holder has one – removed, you'll see three secondary adjustment screws (see Figure 8.2), which are spaced evenly around the periphery of your secondary mirror mount. On most telescopes these will be Allen-head screws, and will require a small Allen wrench for adjustment. A small Allen key should have been included with your telescope, but if not, these tools are very inexpensive and can be purchased at almost any hardware store. The very latest Celestron SCTs have replaced these Allen screws with standard Phillips (crosspoint) screws which are turned with a screwdriver.

How does turning these screws adjust the aim of the mirror located on the other side of the secondary mounting? On most SCTs the secondary mirror is attached to the mounting with a central pivot. Tightening and loosening the screws located around the edge of the mount causes them to push against the mirror, tilting or rocking it in one direction or another. Tightening a screw will, of course, push the mirror against the opposite screw. As long as this screw is not overly tight, tightening its opposite number will move the mirror by the small amounts needed in collimation adjustments.

It should be fairly obvious, looking in the eyepiece at the out-of-focus star, which direction the dark secondary shadow needs to be moved in order to center things up. What may not be obvious is which screw you need to turn in order to move the spot the right way. Rather than scratching your head and trying to figure this out, I suggest you simply choose the expedient of picking one screw and gently tightening it a little. Observe how the dark spot moves, and try the screw's opposite number if it doesn't move in the right direction. If it doesn't seem to move at all, turn it a little more. But resist the temptation to turn the screws by large amounts. The secret to easy collimation is working slowly and methodically.

Figure 8.2. The collimation of your telescope is adjusted by turning the three adjustment screws found on the secondary mounting.

When you've determined which screw or combination of screws you need to turn in order to center the secondary shadow, slowly turn this screw, stopping frequently to peep through the eyepiece, until the shadow is as nearly centered as is possible. *Always adjust your secondary by gently tightening the screws.* Only if a screw is *completely tightened* and can't be easily turned any more should you then loosen the opposite screw to continue movement in the same direction.

When you turn the screws and move the secondary mirror, you'll see that the image of the defocused star moves in the eyepiece field. This is normal. The position of the blob in the field is not critical in this stage of collimation, but you may use the slow-motion controls to recenter it exactly if desired.

Once you've got the secondary shadow centered, stop. Don't tighten down any screws or make any further adjustments. You don't have to lock anything down. As long as you've followed the rule of "never loosen a screw unless you can't tighten its opposite number further," the secondary will be completely secure.

Fine Collimation

You've done a rough collimation using the secondary's shadow. But this just isn't good enough, not if you like to look at planets. Even deep sky observers can benefit from exacting optical alignment. Star images will look tighter and will not feature the "comet tails" so common in the images of miscollimated scopes. To accomplish fine collimation, center Polaris if it's drifted off to the side of the field and, using your focus control, sharpen up the star until you can see a series of diffraction rings around a bright point of light. In the center of the rings, you'll see a small speck of light, the Airy disk of the star. Make the star image large enough that you can make out the rings easily (see Figure 8.3). I generally increase my magnification to about 200× for the fine collimation stage.

Once you can see the bullseye formed by the slightly out of focus star, you should examine it carefully and determine whether the rings are *concentric* or not. Is everything *centered*? Does the combination of Airy disk and rings look like a perfect little bullseye? If yes, you're done. But if the rings seem skewed to one side or

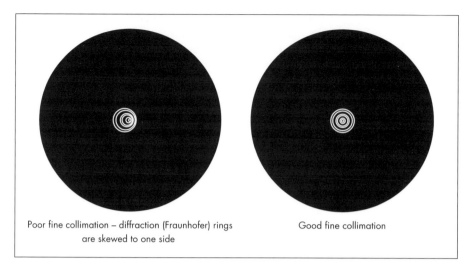

Poor fine collimation – diffraction (Fraunhofer) rings are skewed to one side

Good fine collimation

another, you've got more adjusting to do. Move the secondary screws by *very* small amounts, turning them in even smaller increments than you did when you did your rough collimation. Keep turning screws and checking through the eyepiece until the rings are concentric. Work slowly. There's no hurry. *Always* recenter the star carefully before making a further adjustment. If the star image is off to the side of the field, it may be distorted and falsely indicate mis-collimation when you've got things properly adjusted. Once you have a perfect bullseye, you can relax, your telescope is at least in very good collimation and will give excellent images if it has good optics.

What can go wrong during collimation? If you follow these instructions exactly, there are really only two possible difficulties you may encounter. The most serious is that you may not be able to clearly see the diffraction rings. If you're using sufficient magnifica-tion, you may not be defocused by the proper amount. If you defocus too much, you'll encounter the rough collimation blob again. If you don't defocus *enough*, the rings will not be obvious (if you're exactly in focus, you may see a "real" diffraction ring or two and the Airy disk of the star, but these will be much smaller than the pattern you'll see if you defocused the proper amount). At the right spot, the ring pattern should be around 30 seconds of arc across. If you're defocused but still can't see the rings clearly, the most likely culprit is poor seeing or an uncooled telescope. Poor atmospheric seeing can make the rings hard to see. Unless the

Figure 8.3. Fine collimation is accomplished by making a slightly out of focus star's diffraction rings concentric.

atmosphere is steady, the ring pattern will be moving and smeared, making it impossible to tell when the rings are properly concentric.

What if you just can't seem to get a good night but are anxious to get your CAT critically adjusted? You can use an artificial star. The time-tested star substitute for amateur astronomers doing collimation has always been the reflection of the Sun off a distant power or telephone pole insulator. For best results, try to use an insulator that is at least 25 meters away. Center the reflection in the telescope (taking care not to point your scope anywhere near the *real* Sun), and use the exact same method as above to collimate. If a power pole insulator is not handy, many amateurs have had good luck using the glint of the Sun off a spherical glass Christmas tree ornament.

In-focus Collimation

Figure 8.4. In focus collimation uses the tiny diffraction rings of a sharply focused star.

Once you've finished the fine-tuning stage of collimation, your SCT should be able to take on the most challenging observing assignments. But you *can* go a step further if you're really obsessive about optical performance. The next stage, stage 3, is collimation via the in-focus image of a star. To do this, you'll have to run the power up to at least 300× or maybe even more. You need this much power to reveal the true Airy disk and diffraction ring(s) of a star that is perfectly focused (see Figure 8.4). You'll also need a night of superior

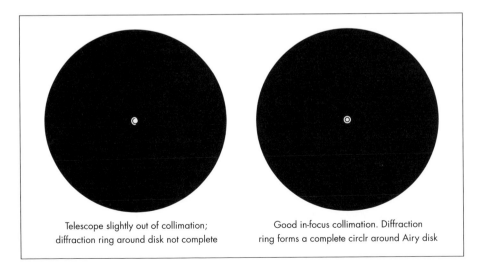

Telescope slightly out of collimation; diffraction ring around disk not complete

Good in-focus collimation. Diffraction ring forms a complete circlr around Airy disk

seeing. The slightest turbulence will make the rings and disk completely invisible, no matter how much magnification you use.

If you're able to clearly see the star's (Polaris is again a good choice) diffraction pattern, a tiny ring or two surrounding the Airy disk as in Figure 8.4, adjust until the ring is concentric around the Airy disk. The appearance of a miscollimated star image is slightly different here than in fine collimation. Instead of being skewed to one side, the diffraction ring or rings will tend to *disappear* for part of their circumference. Adjust the secondary screws by even tinier amounts than you did in fine-tuning until the Airy disk appears surrounded by the diffraction ring. This is really critical work, and you may even find you have to wait for a moment after adjusting the secondary for the heatwaves left in front of the corrector by your hand to dissipate so that you can see the diffraction rings clearly again!

Collimation Tips

Whether you collimate by using in-focus diffraction rings or just leave things at the fine tuning stage, you've entered an exacting realm. Little things can and do make a difference here. One question I'm asked frequently is whether you should collimate the telescope with the star diagonal installed or in "straight-through" configuration with the eyepiece inserted directly into the visual back. There is no doubt a diagonal can affect collimation if its mirror or prism is slightly misaligned. If you adjust the secondary mirror with a diagonal like this in place, your collimation may only be good for one position of the diagonal. If you rotate the diagonal to another viewing position, the prism or mirror errors may mean that your collimation will be off for this new position. More seriously, if you remove the diagonal completely, perhaps to take pictures through the telescope, your scope may then be *out* of collimation.

This subject is slightly controversial among SCT users, but my opinion is that you should collimate with your diagonal in place *if you intend to observe with said diagonal*. If you suspect the star diagonal's optics are misaligned and intend to do mostly "straight through" work – either observing or imaging – remove the star diagonal and do a straight-through collimation. You can tell fairly easily when a diagonal is "off" by rotating it with a star centered in the field of a medium-power eyepiece. If the prism or mirror is not right, the image of the star will move a considerable distance in the field of your eyepiece when you rotate the diagonal. One possible solution to this problem is to purchase a high-quality star diagonal. The stock units that come with new telescopes are now often of fairly low quality.

A common question novices ask is, "How *often* should you perform collimation?" The answer is, "As often as necessary." You need to check the alignment of your telescope optics on a regular basis. This takes little time. When you know what to look for, a quick glance at a barely out of focus star's ring pattern will tell you if your CAT needs attention. A good time to do this is when you've performed a polar alignment by placing Polaris in your field. Before moving on to observing, twist the focuser a bit and take a critical look at the diffraction rings. How often should you expect to find collimation needing adjustment? That depends. If you usually observe from home, your CAT may very rarely need alignment. If, however, you're in the habit of transporting your SCT to distant sites, especially if you have to travel over rough roads, you may have to collimate fairly frequently. But take heart, you at least shouldn't have to adjust every single time like a Dob owner!

Are there any tools that can help you with collimation? Laser collimators designed *specifically* for use with SCTs are now beginning to appear on the market and the initial reports I've had are encouraging. These units require that you be able to perform an exact collimation the "old" star test way first (usually on an artificial star). Once you've got the SCT exactly aligned, you can use the laser collimator to recollimate your telescope from then on, using the position of the laser's projected dot on an engraved grid on the collimator as a reference.

Other Maintenance

Collimation isn't the only work you'll have to perform. There's some other periodic maintenance you'll have to do to keep your CAT in good shape. One caution: before you do *anything* to your new telescope, make sure that you won't be voiding the manufacturer's warranty. The warranty is your insurance policy in case something goes wrong early in the life of your telescope (most manufacturers' warranties are about one year).

Optical Cleaning

When it comes to cleaning telescope optics, SCT users are, as in collimation, lucky. This is because we normally only have one optical surface to clean, the outer side of the corrector plate. Newtonian owners, in contrast, must deal with two optical elements. Our correctors are also much less likely to be scratched by wrong-headed cleaning than are the extremely delicate first-surface mirrors of a Newtonian reflecting telescope. But cleaning any optical device is *not* something to be done lightly. In fact, my usual advice to novices when it comes to cleaning is: "don't!" When cleaning any optical surface the fact is a new user is much more likely to do more *harm* than *good*. Does that mean you should never clean your corrector? No. If it is really dirty and in need of cleaning, you can proceed. But many novices obsess about cleanliness of their SCTs.

The truth is, no matter *what* you do, the corrector plate of your telescope will *always* have some dust on it. This is inevitable if you actually use your telescope and it will not harm a thing. Don't worry about it, and don't be tempted to over-clean. A speck of dust will do much less harm than a big scratch you've left on the lens by mistake during one cleaning too many. This really can't be overemphasized: *leave your corrector alone unless it's really, really dirty*. It would do many new amateurs good to see how dirty the optics of professional telescopes are allowed to become before they are cleaned.

But the time will eventually come when the SCT or MCT *will* need a gentle corrector cleaning. In addition to a little dust, it may have accumulated a finger print smudge or two. Or some unidentifiable gunk was deposited on it at that last star party. If it really is obviously dirty, there's no denying it's lens-cleaning time. What's the secret to preserving your beautiful corrector lens and getting it nice and clean? The secret is to treat your corrector just as you would a fine camera lens – which is really what it is. It is precisely aligned and multicoated for optimum performance. You should treat it with due respect, but cleaning your corrector needn't be a frightening prospect. I'm just attempting to impress upon you the delicate nature of all optics and the need for exercising extreme care when you're cleaning them.

Corrector Cleaning

Before outlining the proper procedure for cleaning optics, let's discuss for a moment how you can keep your optics clean enough so as to minimize the need for cleaning. You should always keep your corrector lens cap in place when the scope is not being used. Even when the scope is outside cooling down prior to use, it's a good idea to keep the cover on. I also feel that storing the telescope in a case or box also tends to minimize contamination. That SCT may look pretty assembled on its tripod in your living room, but a case of some kind will keep it both cleaner and safer. It's also wise to keep the rear port of the telescope closed at all times. Dust, bugs and other contaminants can enter your OTA, and removing these foreign substances is a major job.

A major source of contamination for your corrector plate is dew. A wet corrector can cause any dust present to adhere to the lens. It can also promote the growth of fungi. Keeping dew from adding to your cleanliness problems is simple enough, though. If you live in an environment where heavy dew formation is likely, you'll have to equip yourself with dew fighting tools and keep your corrector dry if you're going to do much observing at all. What if, however, in spite of your best efforts, your corrector is soaked in dew by the end of the evening? This isn't a disaster as long as you keep the lens cap *off*. Never, and I do mean *never*, cap any dew-covered surface. If you do, you're likely to promote fungal growth as the capped lens slowly dries. The dew may also leave behind a deposit of some kind. If your location is humid, the scope may *never* dry completely. Remove the dew with a heater gun before capping the telescope aperture or move the CAT inside and let it dry naturally outside its case. To lessen the chance of dust collection, point the corrector end down. Once the lens is completely dry, you can cover the corrector and return the SCT to its case as normal.

Cleaning Procedure

Step one in cleaning is to deal with the easy stuff first, the loose accumulation of dust lying on the corrector's surface. The most efficacious method for getting rid of these particles is with canned air. You've no doubt seen cans of compressed air designed for cleaning electronics and photographic equipment for sale in various locations. Be sure to purchase a brand designed expressly for photo/optical use, as it is more likely to be free of possibly damaging contaminants than that designed for cleaning computer circuit boards or other items. Generally, any name brand sold in a camera store should be fine.

To use canned air to clean your SCT's corrector plate, move the OTA in declination until it's level and lock it down. Remove your lens cap and, holding the canned air gun absolutely level; give your corrector a few blasts until the loose dust is blown away. Don't hold down the trigger, instead apply 2–4 second bursts. You may find it helpful to point the air blast at the corrector at an angle during the dusting process for best results. Continue blasting the plate with air until all of the loose dust has been blown away. Take care to keep the can level to prevent any liquid propellant from being expelled onto your lens where it might potentially do damage (due to its chemical composition or freezing cold temperature). For safety's sake, also keep the air nozzle at least a foot away from the surface of the corrector.

Some amateur astronomers fear canned air, having been told by those in the know that you should never use it around telescopes. Like many stories passed from amateur to amateur over the Internet or in the form of club gossip, there is a grain of truth in this. It is inadvisable to use canned air on a first surface mirror. Depositing propellant on a mirror or blowing a particle of foreign matter on it can do real damage. But one of the major applications for canned air *is* cleaning lenses, which is what your corrector plate is, and it works very well in this application if the above safety precautions are observed.

If you don't want to use canned air, either because you don't trust it or can't find it in a photographic grade, a good substitute is a blower brush. This is a little plastic item sold in just about every camera store I've ever visited. It combines a small bulb-type squeeze blower with a brush. The air blows through the brush when the bulb is squeezed. If your corrector is only a little dusty, this can be a good tool. I do advise that you limit your use of this implement to just the blower portion. The brush part on every example I've seen looks a little dirty right out of the box, and these brushes do not seem to be made of camel's hair, which is the only sort of brush that should ever touch a lens.

What if some stubborn dust remains on the corrector after your canned air or blower brush dusting job? Or if, in addition to loose dust, your corrector has smudges and spots on its surface? You proceed to the next level of cleaning. For this you use lens tissues, either alone or in concert with lens-cleaning fluid. To obtain lens tissue and fluid visit a reputable camera store just as you did in your purchase of canned air. *Never* use the lens tissues designed for cleaning eyeglass lenses. These are not only often of questionable manufacture, they are also, in many cases, impregnated with a silicone compound that will have the effect of making your corrector plate dirtier than it was before you cleaned it. The same thing goes for the lens-cleaning fluid sold in places other than photo shops. Who knows what's in it? It is just as likely to make matters worse than better, from what I've seen.

At the camera store, resist the efforts of the staff to sell you "reusable" lens-cleaning cloths. These soft cloths are used for a while and then laundered to return them to a clean state. The problem is knowing when to clean the cloth and whether laundering always removes grit and other fatal scratch-causing contaminants. What you want is disposable paper lens tissue sold for cleaning camera lenses – only. I favor the Kodak brand (sold in little envelopes of the world-famous Kodak orange/yellow color). It works well and is very inexpensive. Kodak also makes a very good grade of lens-cleaning fluid.

Once you have your tissue and fluid, you can deal with your corrector's remaining problems. For stubborn dust, you can use a tissue to gently *brush* the surface of the corrector plate. *Don't* apply pressure, just hold a tissue and sweep it across the surface until the dust is history. Change tissues frequently during this process. Lens tissue is cheap; your corrector is expensive. What about more stubborn dust and those small smudges and oily deposits? If they're not too toughly adhering, you can use breath condensation and a tissue to remove them. Breathe lightly on the corrector until the area you want to clean is fogged and gently, very gently, wipe the area with a tissue until the condensation and the underlying dirt are gone. Use linear strokes, never circular ones, and use different areas of the tissue with every single stroke. Once the tissue is used up, get a new one. This is very important. And always wipe from the center of the corrector toward the edge.

Should some spots prove to be very stubborn, you may bring your lens-cleaning fluid into play. Always apply the fluid *to the tissue* and not to the surface of the corrector lens. It's all too easy, if you drop the fluid onto the lens, to use too much. You then run the risk of some spilling to the edge of the corrector and migrating into the interior of your OTA where it can leave stains or promote fungal growth as it slowly dries. When using fluid, clean just as you did with breath condensation, changing tissues frequently and using linear strokes from the center of the lens outward. Continue gently cleaning with minimum pressure and changing tissues until the lens fluid left on the plate is dry and all dust and gunk are gone. In most cases this is all you have to do. Your corrector plate is clean and shouldn't need another cleaning for quite some time. Dispose of the used lens tissues, cap your lens-cleaning fluid and put the aperture cover back on the telescope.

These cleaning methods should work well for all normal optics cleaning tasks and are applicable to eyepieces and finder objectives as well as to corrector plates. There are times when something more is needed, however. If your corrector is really filthy dirty and is covered by large dirty areas, you'll need a heavier-duty method of cleaning and a "stronger" cleaning fluid. How could a telescope get that dirty? You'll see this level of contamination on used instruments that have been stored without being capped. Anything can happen, though. One touch of a child's greasy fingers can leave you with a badly smudged lens (I've seen this happen more than once at public star parties). Most commercial lens-cleaning fluids are nothing more than distilled water with perhaps a small amount of detergent-like wetting agent added. This works fine for the small jobs, but for heavy grease and oil you're advised to make your own cleaning fluid.

The ingredients needed for your "super" lens-cleaning fluid are few and are easily available. Start with isopropyl alcohol. This is the denatured alcohol, available from chemical suppliers and pharmacists. Don't use rubbing alcohol, as it contains water of unknown quality and often impurities as well. You'll also need distilled water. Not *bottled* water, *distilled* water. A final ingredient is dishwashing liquid. By this, I mean hand dishwashing liquid, not something that goes in a dishwashing machine. Try to use a brand labeled as biodegradable. The recipe for the fluid is also a simple one: add two parts distilled water to one part isopropyl alcohol. To this mixture, add one drop of dishwashing detergent per quart of liquid.

Once again, apply the cleaning fluid *only* to the tissues, never to the surface of your corrector plate. If the areas to be cleaned are really large, you may, if you wish, use two or three tissues at the same time. You should still rotate the tissues to expose unused areas with every stroke, change tissues frequently, and swab gently in straight strokes from center outward. Continue cleaning, wetting fresh tissues with fluid, until all dirt has been removed from the surface. Dry any remaining moisture very gently, dabbing with clean tissues.

Cleaning the Inside of an SCT Corrector Plate

Under normal conditions you may never have to worry about opening up your SCT in order to clean the inside surface of the corrector plate or the surfaces of your primary and secondary mirrors. An SCT that is kept in a case with the rear port covered will likely go many years – perhaps even a lifetime – before "inside" problems happen. But in some cases dust can eventually build up in enough quantity on the optical surfaces inside the OTA to cause performance problems. An SCT which is stored unused for a long time can also be prone to exhibit problems like a hazy film over the inside of the corrector plate caused by the outgassing of materials in the OTA of an unused instrument stored in a warm environment.

What do you do when a problem develops with an interior optical surface of your telescope? The first thing to consider is whether you actually *have* a problem or not. As was discussed in the chapter on initial checkout of the SCT, a flashlight held at just the right angle and pointed down your OTA can cause even a few specks of dust to stand out in sharp relief, making your primary mirror look absolutely *horrible*. Don't worry about a little dust; it won't hurt a thing on your mirror or on the inside of your corrector plate. If, however, your primary or secondary mirrors have developed a *really* thick coating of dust over the years or if you notice that your corrector plate doesn't look exactly clear anymore, you do need to consider taking action.

The best course of action for many people will be to simply ship the scope back to the manufacturer for cleaning, especially if dirty mirrors are the problem. These problems with interior cleanliness are likely to develop with older telescopes. They are good candidates for cleaning of not only optical parts but mechanical ones as well and – quite importantly for older SCTs and moving mirror MCTs – relubing of the focusing mechanism. You really should depend on the manufacturer for these tasks, which may involve total disassembly of the OTA and drive base. The problems with sending the CAT to the manufacturer are that you'll be without it for a while, you'll have to trust it to the tender mercies of the motor freight folk, and the bill, even just for an optical cleaning, can be fairly substantial when coupled with freight charges. If the primary and secondary mirrors require attention, I would certainly advise having a professional look at the scope, though.

Cleaning the Inside of the Corrector Plate

If you decide you don't want to or can't afford to ship your SCT back to the manufacturer and all it needs is for the inside of the corrector to be cleaned, you can undertake interior cleaning yourself (if you live in a large city, there *may* be an astronomy dealer who can do the job for you). This is a perfectly reasonable course of action, as long as you understand a few basic facts. Foremost is that delving into the interior of your telescope's OTA will most likely void its warranty. Also, any mistake you make that you can't correct yourself, ranging from accidental misalignment of optics to a worst case scenario of breaking your corrector plate, will likely cost you more in repairs than a cleaning by the maker would have. Additionally, I advise MCT owners to leave removal of their large correctors to the manufacturer and to *always* return the scope to the factory if corrector plate removal is needed for any reason. *If, and only if,* you accept *full responsibility* for these possible outcomes should you proceed to attempt a cleaning of your telescope's interior optical components as described in the following paragraphs.

After hearing all these dire warnings, the actual act of removing the corrector plate to access its inside surface for cleaning seems rather anticlimactic. There are a couple of precautions to observe, however. Most important you *must* maintain the same *radial orientation* of your corrector when you replace it. It must go back in exactly the same position in regards to rotation as it was originally. If you accidentally rotate the corrector when you reassemble it, images will become badly degraded in many cases. Many Meade telescopes have marks on the corrector and the backing ring behind it that holds it in place to indicate proper corrector plate orientation. If you don't see any obvious marks, clearly mark both corrector and tube interior with a soft pencil so you can replace the lens the correct way. Some telescopes may also have small spacers that serve to tilt the corrector one way or another. If you notice any little shims, try not to disturb them, and mark their positions so they can be replaced during reassembly if they become dislodged. The shims may take the form of small pieces of cork or paper. It would be a very good idea, when removing a corrector, to have a pencil and notepad at hand to make notes on disassembly as an aid in putting things back together.

Before beginning removal, ensure that the SCT is on a solid, clean surface. Tilt the scope up at a 45 degree angle so the corrector won't fall out when its retaining ring is removed. With both Meade and Celestron telescopes you remove the corrector plate by removing the screws (either Allen or Phillips head types) in the plastic retainer ring on the front of the scope. Once this ring is removed, check for the presence of spacers, as they may only be evident with it out of the way. The plastic ring is all that holds the lens in place. With the ring off and laid aside in a safe place along with its screws, you may remove the corrector plate itself by grasping it by the secondary mounting and gently pulling. When the corrector comes free (you may have to rock it from side to side a bit to get it moving), set it aside, inside surface up (to protect the secondary mirror) on a padded surface like a nice clean towel. Handle the corrector only by its edges or by the secondary mounting.

You can now clean its inside surface just as you did the outside, using tissues, cleaning fluid and canned air. Just be careful. The corrector plate is almost window glass thin and similarly easy to break. Replacing this precision optical element will cost substantially more than replacing a broken window, though! And keep your fingers away from the exposed surface of your secondary mirror – it is somewhat protected by the short baffle that extends out from its mounting. With the corrector's insides nice and squeaky clean, you can proceed to reinstall the lens. Before setting it into the OTA, you must, of course, replace shims – if any – in their original positions. With these spacers in their appointed places, tilt the corrector on edge on its pad, lifting it by its edges only, and, holding it in place with one hand, grasp the outer side of the secondary mount *firmly*. Gently lower the corrector onto the backing ring, observing the proper radial orientation using your reference marks. You then replace the plastic retaining ring, taking all due care not to shift the corrector's position. You'll probably find that you've managed to leave a few fingerprints on the outer surface of the corrector during the process of replacing the retainer. If so, clean them with our usual method. You're done! Congratulations. The task of removing the corrector is not one to be undertaken lightly, but if the procedure given here is followed, it really is a pretty easy operation.

Star Testing for Optical Quality

Your optics are clean, but how good are they? Amateurs often say they have "good" optics in their telescopes. But usually their opinions as to how well their telescopes perform are based on subjective criteria. The Moon looks good. You can see Cassini's Division on Saturn. Stars seem pretty small even at fairly high magnifications. Things like these actually *can* indicate you have good optics. We can even develop some general guidelines as to how a planet, for example, should look if your optics are up to snuff. But visual impressions are subjective and open to quite a bit of interpretation. Isn't there a more quantifiable method of determining whether you have got a good telescope or not? There certainly is: star testing.

If you used the out-of-focus image of a star to help you collimate your telescope, you've already performed a type of star test. You used the arrangement of the diffraction rings of a barely out-of-focus star to gauge your scope's alignment and to help you fix any miscollimation. We can also use these diffraction, or Fraunhofer rings, to tell us a lot about the quality of our telescope. This time, we'll be interested not in whether the rings are concentric or not, but the general appearance of these rings.

The star test is both extremely simple and amazingly complex. It is simple in that we're only looking for one thing: whether the out-of-focus diffraction rings of a star are identical in appearance both inside and outside of focus. When you go from just "outside" focus to just "inside" focus, do the rings appear absolutely identical (see Figure 8.5)? If they do, you can rest assured that your telescope's optics are as close to perfect as can be produced! If not? Well, that's the time when things get a little more involved. Star tests can tell you a great deal about the conditions of your primary and secondary mirrors and corrector plate. The problem comes in evaluating the subtle differences between inside and outside of focus diffraction ring patterns. Each optical problem you confront will make the in/out of focus rings appear subtly different. Some scopes may even exhibit combinations of problems. Luckily, there are really only a few problems users of commercial CATs are likely to encounter, and they are pretty straightforward to diagnose.

The starting point for the star test is choosing a star. Having a good star is more critical for star testing than it is for alignment. A second magnitude star is about right for testing an 8 inch SCT. A star either too bright or too dim for your aperture will make subtle differences in ring appearance unclear. For scopes larger than 8 inches you'll want a star that is dimmer than magnitude 2; for smaller telescopes you'll need a brighter sparkler. The key to knowing whether you've chosen a good alignment star is whether the diffraction rings are very clear and well defined. The atmospheric seeing will have to be steady and your scope will have to be completely cooled down for you to get a good look at those rings, of course.

$200\times$ is about the minimum magnification needed for star testing. While you can possibly get by with less, most authorities suggest using a magnification of at *least* $25\times$ per inch of aperture. The important thing is that you can see the out-of-focus rings clearly.

The first step in the star test is to examine the star's rings to see if they appear concentric. For the test to be decipherable, your optics must be critically aligned. If you don't have a perfect bullseye, adjust your secondary

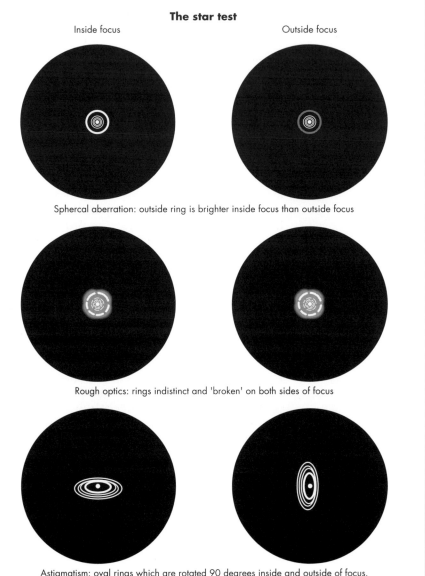

The star test

Inside focus   Outside focus

Sphercal aberration: outside ring is brighter inside focus than outside focus

Rough optics: rings indistinct and 'broken' on both sides of focus

Astigmatism: oval rings which are rotated 90 degrees inside and outside of focus.
Alignment may not be exactly as shown

The star test requires very steady seeing, a completely cooled down telescope and high magnification

Figure 8.5.   Interpreting a star's diffraction rings is the key to a star test for optical quality.

as outlined previously until they are as concentric as you can make them. If the seeing appears poor, with the rings wavering or jumping around, you might as well pack it in for the evening. Even if you can manage to get a decent collimation in conditions like these, bad seeing will make star test interpretation impossible.

If you live in an area where seeing is seldom good, you can use an artificial star rather than a real one for your testing. You'll need a pinpoint light source, something that's easily provided by the Sun glinting off a spherical Christmas tree ornament as suggested in the collimation section. There is one difference when you're performing an optical quality test, however. Your ornament/star *must* be a certain minimum distance from the scope. Experts recommend at least 20 times the focal length of your CAT. This amounts to 41 meters (133 feet) for an 8 inch f/10 telescope (8 × 10 × 20)) – a considerable distance for many city dwellers. But it is necessary. The result of placing your artificial star too close to the telescope will be the false appearance of spherical aberration – a serious defect.

The two aspects of the diffraction rings we'll be examining are their appearance inside focus and outside focus. Inside focus with a Newtonian or a refractor is when your eyepiece is moved past the focus point toward the secondary mirror or objective. When you're outside, you've moved past the focus point away from the secondary mirror. Which is inside and which is outside is easy to determine with these scopes, but is not so obvious with an SCT or a moving mirror MCT. Inside focus with one of these scopes is when you move the mirror *down* the tube, *away from the corrector plate*, from perfect focus. Inside is when you move the mirror up the tube from the point of perfect focus, *toward the corrector*. You need to find out which way the mirror moves when turning the focus control. You can determine your mirror's movement by observing which way it moves, up or down the tube, when you turn the focusing knob one way or the other. It's easiest to figure this out by the light of day or inside the house. You'll have to look closely, as the mirror moves slowly when you turn the control and doesn't move very far in either direction. One tip is that you should feel more tension on the control – especially with a Celestron scope – when moving the mirror toward the corrector.

When you know which way to turn the focus control to move inside or outside of focus, you can proceed. Examine the unfocused diffraction rings of your star carefully – it doesn't matter whether you begin inside or outside of focus. Try to get an image of them locked in your mind. When you're familiar with the way they look, turn your focus control until the star sharpens into focus and keep turning until the diffraction rings reappear on the other side of focus and are approximately the same size as the other set of rings. How do they look? If they look the same, as in Figure 8.5, you've got a perfect telescope! The rings do *not* have to look identical to those shown in my drawings; the only requirement is that the inside/outside focus rings look *the same*.

If the inside and outside of focus rings are not identical, this indicates your telescope suffers from one or more optical aberrations – problems with the figure of the primary mirror, the secondary mirror, the corrector plate or all three elements. But don't panic! Please keep in mind the star test is *incredibly* sensitive, and problems shown up in the test may be minor to the point of not causing any problems at all. But if you do have in/out of focus rings that are not identical in appearance, you should proceed to identify what problem, exactly, your telescope's optics suffer from.

Spherical Aberration

A sphere-shaped mirror cannot form a perfect image. Like the primary in the misfigured Hubble Space Telescope, spherical mirrors do not deliver all light rays to the same focus point. Spherical aberration is a common problem with Schmidt–Cassegrains. They, after all, use a *spherical shaped mirror* as an objective. That special lens on the front of your telescope, the corrector plate, is supposed to remove this spherical aberration. Usually it does a good job of this. But it is a complex optical element, always difficult to manufacture, and can be rendered less than effective not only by errors in its figure but by radial misalignment or tilt. Spherical aberration should *not* be present in large amounts in your SCT's optical system. If it is, it will definitely degrade your telescope's performance, especially in high-resolution applications like planetary observing and image.

What indicates your CAT suffers from this affliction? The tip-off to spherical aberration is that the *outer diffraction ring* appears significantly brighter inside focus than it does outside focus. The inside-focus ring also appears sharper in an instrument with a considerable degree of spherical aberration. A scope that shows strong indications of spherical aberration as in Figure 8.5 would be seriously compromised. One bearing a smaller amount would likely perform well. You can tell, just by looking at the diffraction patterns, that your scope has a problem in this area, but deciding on how bad the problem is may take a set of experienced eyes. A long-time amateur from your club should be able to help you make the determination. What if you can't find help? In that case, a logical next step is to evaluate the scope's actual imaging performance as outlined at the end of this section.

Is spherical aberration the end? Should your telescope be returned? Is it useless? Not necessarily. If I were to see a relatively small amount of spherical aberration in my SCT, I wouldn't let it bother me for a moment. The fact is that many – if not most – SCTs will show *some* spherical aberration due to their focusing method. The corrector on your telescope can only provide *perfect* correction for spherical aberration for one spacing of the optical components. Moving the mirror to focus for a particular eyepiece may move you out of this good zone, and allow small amounts of spherical aberration to remain. If planetary images are good, I wouldn't worry about it.

Rough Optics

Ideally, the surfaces of your telescope's optical components should be polished to a very smooth finish. Unfortunately, this doesn't always happen. Rough optics in commercial telescopes are caused by machine polishing of mirrors. The mass production methods used to produce SCTs make them quite vulnerable to this problem. What is the result of having rough-surface optics? Light is scattered in the optical system. Some contrast is lost and tiny low-contrast planetary details may be completely invisible. Very dim deep sky objects, especially those located adjacent to brighter ones, may also be hard to see.

How do rough optics change the look of the diffraction rings during the star test? In Figure 8.5, you can see the rings actually take on a rough and uneven looking character. They seem a little ragged both inside and outside of focus. The patterns indicating surface roughness will be almost impossible to see except during times of the *very steadiest* seeing. Wait for an exceptional night or use an artificial star during the day if you wish to check for this condition.

Are rough optics a serious problem in current CATs? Not in my opinion. Most *smaller* SCT mirrors produced these days have fairly smooth surfaces. If you find this problem, it's most likely, given what I've seen over the years, to show up in the largest CATs, 14–16 inchers. Even in these big mirrors, the problem seems less prevalent than it used to be. While rough optics can cause problems for the critical observer, I consider this problem to be much less serious than spherical aberration. Given the choice between spherical aberration problems and rough optics, I'll take the rough optics every day!

Astigmatism

Astigmatism is an optical aberration, which, while as serious as spherical aberration, is, thankfully, rare in Schmidt–Cassegrain telescopes. Astigmatism is the result of problems in the primary mirror's original blank – stresses that caused the figure to be rendered malformed when the mirror was ground and polished. Astigmatism, once you know what to look for, is completely unmistakable. It appears as in Figure 8.5. On one side of the point of focus your diffraction rings appear elliptical. As you pass through focus to the other side, the elliptical rings change their orientation until the ellipse's long end, its major axis, is oriented 90 degrees from the original position.

Yes, astigmatism is rare in SCTs. But it is not completely unheard of. This aberration can result from problems with the corrector and secondary mirror as well as from problems with your primary. I've occasionally seen astigmatism caused by too-tight collimation screws on the secondary mirror mount. Get them too tight and they can cause a warping of the mirror, resulting in astigmatism. Don't be fooled into thinking your telescope suffers from astigmatism if it doesn't. The appearance of astigmatism can easily be caused by an eyepiece or your eyes! If you think the eyepiece you're using is astigmatic, the easiest check is just to try another of a similar focal length. If you suspect the fault is in *you*, having another observer take a peek should reveal this.

Other Aberrations

Turned-down edge can also ruin a telescope's performance, being nearly as fatal as spherical aberration or astigmatism. It is the result of the outside edge of the mirror being ground so it slopes or rolls off at an angle opposite to that of the mirror's curve. It's as if you took a file and beveled the edge of your mirror's surface so it no longer ends sharply. The star test easily reveals turned down edge. Inside focus, the diffraction rings are very sharp and contrasty; outside focus they soften, blur, and almost disappear. Like rough optics, turned-down edge is a contrast-killer. Luckily it's not often found in today's CATs, being seen mainly in homemade Newtonian mirrors.

And there are other possible optics problems as well. Various *zonal errors* – hills and valleys on portions of your mirror's surface – can be present. These are detectable with the star test, but you probably won't find them as they are rarely seen in SCTs. If you'd like to learn more about star testing and become proficient with it to a greater degree than just being able to detect gross amounts of spherical aberration, I recommend you track down a copy of Harold Richard Suiter's wonderful book, *Star Testing Astronomical Telescopes*. This little volume will lead you through all the nuances of the star testing procedure and will teach you a great deal about telescope optics in general.

Planetary Image Evaluation

Say you detect a bit of spherical aberration or other problem in your optics. Is it enough to cause a problem? You can answer this question by giving the images you're seeing a frank, honest evaluation. The most telling object to focus on is a planet – Jupiter, Saturn or Mars – which should display a decent amount of detail to a good telescope. Ideally, you could set your telescope up next to another CAT of the same aperture that is known to have good optics. If this is not possible, use the following summaries of what an 8 inch telescope should resolve. Smaller scopes may show a little less detail and bigger ones may, if the seeing is right, show a little more. Seeing is very important for this type of evaluation. If it's not rock solid, your scope cannot and will not deliver the detail it is theoretically capable of providing. The following assumes that your scope is perfectly collimated, too.

Jupiter

At 200×, the image should be sharp, though Jupiter *is* a little soft-edged and will not appear quite as clearly defined as Saturn and its rings. There shouldn't be a whole lot of question as to when the telescope is in focus, though. It should snap into focus. The Galilean moons, under the best seeing conditions, should appear as minute disks of subtly different sizes. Even a cursory glance should show that they are not stars. What about Jupiter's surface? How many cloud bands you can pick out will depend on the planet's current weather patterns, but if all you can see are the north and south equatorial bands, something is not right with your scope. In addition to being able to make out additional and more subtle bands, you should see some *detail* in the bands. At a minimum, you should be able to detect that the edges are not straight, but irregular. Projections like festoons should also be visible if present.

Saturn

At 200–300×, Saturn should show even more focus snap than Jupiter. It should hold up well to higher magnifications on good, steady nights. The rings, when tilted much past the edge-on aspect, should show the Cassini Division *easily* as a razor-thin sharp black line. You should also notice brightness differences across the A and B rings. When seeing's good and the rings are anywhere close to wide open you should be able to see a hint of the dusky Crepe ring. Saturn's surface features are subtle, but you should be able to at least see some banding – easily – you shouldn't have to guess at it. In addition to Titan, two or three of Saturn's other moons should be "easy."

Mars

When the planet is at opposition you should be able to make out dark markings (unless Mars is turning one of its less detailed faces to Earth on the night you choose to observe it). A polar cap should be apparent, even when shrunken by the Martian spring. Mars is small, even when it's relatively close, but your scope should allow you to pour on enough power, say about 300× on steady evenings, to make it easy to pick out surface detail.

If your scope has shown up an obvious optical problem in the star test and will not display the level of details described above *on a fine night*, its performance is definitely below par. Don't dismiss a lack of performance as being the fault of the SCT design. These scopes, when well made, are *easily* capable of showing you this kind of detail and more. There is one caveat here: if you are a new observer you may not be able to see much on the planets with any telescope, no matter how good its optics. Seeing fine features on Earth's sister worlds requires some experience. If you're not a planetary veteran, enlist the services of a local planetary observer, preferably one experienced in SCT use. Ask for a frank and fair opinion on how things look in your SCT.

Your scope shows "bad" in the star test and your planetary images reflect this, being so-so at best. Just to be sure, you've redone the star test on a night when you're *sure* the seeing was good. You've double-checked collimation, getting help if necessary to touch up your alignment. And you've also gotten the advice of a local experienced amateur. Sadly, he's given that new scope a thumbs-down. What do you do? If this is a new telescope, contact the manufacturer. Both of the SCT makers have remarkably good track records when it comes to dealing with poor optics. Just be sure you have your facts straight. Make notes about how you performed the star test, what you saw in diffraction patterns, and, perhaps most importantly, how the telescope is performing on real-world objects like the planets. I think you'll find both Meade and Celestron willing and even eager to help you resolve your problem.

If this is a used scope, I'm afraid your options are very limited. What you *should* do is perform the star test, or, at least, a planetary detail check *before* you buy a preowned scope. If you failed to do this and forked over your money, don't be surprised if the attitude of the seller is, "You bought it, it's yours." Can you do anything other than just grin and bear it? Well, you can go to the manufacturer, but it's likely that any fix will cost real money. Of course, that's better than being saddled with a poor scope. If the optics are really bad, you will be tempted not to use the scope much and may even lose interest in astronomy. But you should also keep in mind that you can do much good observing, even with pretty poor optics. If your prime interest is deep sky visual observing and imaging, you may not even notice any difference between your "lemon" and a superb OTA.

Aligning Declination Setting Circles

If you're to have a prayer of finding objects using your telescope's analog setting circles, they must be precisely calibrated. The RA circle must be set every time you use the telescope by dialing in the right ascension of a bright star currently in the scope's high power field of view. The declination setting circles, on the other hand, are set at the factory and usually will not need user adjustment – at least not very often. Over time, these circles can fall out of adjustment, thwarting your best efforts to dial in celestial sights. Correctly setting up the circle is not difficult.

The only items you'll need to calibrate your declination circle are a spirit (bubble) level and a screwdriver or wrench to loosen the declination circle's lock and allow its scale to be turned freely. Pick a nice, level, solid place on which to set your telescope. A sturdy table or a level concrete floor will work well. Don't drag out the tripod; just set the scope's drive base on this level surface. Ensure that the drive base is absolutely level using your level. Place this tool across the length and width of your scope's base checking the bubble to see that it is centered. If not, the base is tilted. Use shims under the base to adjust its position until the level indicates no tilt in any direction. Small pieces of cardboard and paper can be used under the base to adjust it.

When you're sure the base of the telescope is level, you can proceed to check the declination circle for accurate calibration. Loosen the declination lock and set the scope to a declination of 90 degrees. If your telescope has two declination circles choose the one adjacent to the declination lock for making the setting. Pointing your wedgeless scope at 90 will result in the scope's tube pointing straight up. Take care to align the

Figure 8.6. A bubble level is used to calibrate the CAT's declination setting circle.

"90" reading on your circle as closely as possible with the pointer. When you're satisfied, lock the declination lock securely, taking care not to move the scope in declination. Then, remove your telescope's lens cap and, being careful not to touch the corrector with your fingers or your level, place the level across the end of your tube perpendicular to the fork arms as in Figure 8.6.

Take a look at your level's indicator. Is the bubble precisely centered in the marks? If not, turn your declination slow-motion control one way or the other until the bubble is centered. When you're happy with the bubble's position, loosen the right ascension (not declination) lock, and, without moving the level, rotate the tube 90 degrees. Observe the bubble's position again. If it is not centered, turn the declination slow-motion control until it is. If you have to do this, rotate the scope back 90 degrees to its initial position and make sure that the bubble is still centered. If you can't seem to keep the bubble centered when you move the scope 90 degrees, use your level to check the base. It may be tilted after all.

When you're sure you've got the tube assembly absolutely level, you can check your declination setting circles. Take a close look at the pointer. Is it *precisely* positioned on the 90 degree mark? If you had to turn the declination slow-motion control in order to center the bubble in the level, it probably isn't any more. To correct this, loosen your declination circle's lock. Refer to your telescope's instruction manual to find out how to do this. Most Meade telescopes have a knob in the center of the declination circle that is turned counter-clockwise to loosen it. The scale of the setting circle can then be turned freely in order to calibrate it. Celestron SCTs require that you remove a cover from the declination bearing assembly and loosen a screw or bolt centered on the declination axis.

Without moving the scope in declination, turn the scale until it indicates 90 at the pointer. Retighten the knob or bolt that secures the circle very carefully, holding the setting circle with one hand so you don't accidentally move it while locking the scale back down. Don't force anything, of course, but lock the circle down firmly so it's not likely to wander from its now calibrated position easily. If your telescope has two setting circles, you may now perform exactly the same procedure on the other one. Precisely aligning your declination setting circle in this fashion will not only aid you in object location, but will help speed up your polar alignments.

Troubleshooting

Things *can* go wrong with your telescope. Any complex mechanical, optical and electrical assemblage like a modern SCT will probably suffer some sort of problem at some point. The troubleshooting chart in Table 8.1 is provided as a general guide to SCT problem diagnosis and correction. Perhaps the greatest value of this for the beginner will be in helping you determine whether there's something really wrong with your telescope or whether fuzzy images are merely the result, for example, of unsteady seeing or a need for some simple collimation adjustments. Naturally, not every problem or problem condition could be covered here. Always use your telescope's owner's manual as your basic guide, especially when you encounter difficulties with a sophisticated computerized telescope.

Contacting Telescope Makers for Assistance

If, after reviewing your manual and the troubleshooting charts in Table 8.1, you think your SCT has a problem that requires assistance from the manufacturer, don't hesitate to call for help. Do make sure you have all your facts straight – you're positive about your scope's symptoms and you've exhausted simple troubleshooting procedures. It would be embarrassing to return a scope to the manufacturer only to have the technicians determine your battery was dead! Power problems are probably the number one cause of electromechanical difficulties with modern scopes. Check and recheck batteries and power connections, especially if you're operating on battery power on a cold night. Before you call, be sure you have the data on your scope's serial number, exact model number and purchase date/location at hand.

As far as actually dealing with support technicians, the best advice I can give you is "keep your cool." You may well be very upset to find you have a defective telescope, especially one that is fairly new (for problems with a *brand new* scope, contact the *seller* first). But you'll accomplish far more by talking calmly and rationally and explaining your problem. Follow any instructions you're given carefully, writing them down and reviewing them with the technician before you ring off. Do stand your ground, too. If you're *certain* your scope has a real problem, don't let yourself be brushed off. Also be aware of your rights in regard to your warranty and exercise them.

Table 8.1.

Symptom	Probable Cause	Corrective Action
Optical/Image Problems		
Images of stars, planets and other objects are not sharp and clear. Image cannot be properly focused. Planetary detail lacking.	Poor seeing conditions, indicated by an image that "boils", wavers or moves around in the field of view, especially at high magnifications. The air is unsteady due to atmospheric conditions or because the object being observed is too close to the horizon. Objects close to the horizon will display color effects in addition to boiling and image movements. Poor seeing can also be caused by local terrain and buildings. Observing on or across a roof or parking lot radiating heat will result in very poor "seeing".	Wait until seeing conditions improve. Wait for target object to rise at least 30 degrees above the horizon, or catch it earlier in the evening before it sinks in the west.
	Telescope has not cooled sufficiently. If the telescope optics have not had time to assume ambient temperature after being moved outdoors, images will be poor, resembling those produced by poor atmospheric seeing. A warm telescope will also display tube currents, cells of warm air circulating in the telescope which will cause bad "seeing".	Allow telescope to cool sufficiently. If the temperature differential between inside and outside is small, 30 minutes is adequate cool-down time. If, however, the temperature difference is large or critical, high-resolution observing or imaging is to be attempted, at least 2 hours should be allowed.
	Image not properly focused.	Turn telescope focus control until image is sharp. Turn control in direction which tends to make image "smaller". Continue turning through best focus until image begins to unfocus again, and then turn in opposite direction to bring image into maximum clarity. Turn the control through and back to best focus several times until you're sure the image is as sharp as possible. If the image is considerably out of focus initially, bright objects will appear as large, dim globes of light with dark centers (the shadow of the secondary mirror). If telescope is very much out of focus all but the

	brightest objects may be invisible. Point scope at a bright star, or planet or a distant terrestrial object like a streetlight, and turn focuser until the target begins to get smaller. If no effect, or the opposite effect is seen, turn focuser in the opposite direction.
Your eyeglasses are causing problems.	Eyeglasses should be removed for observing through a telescope unless the wearer suffers from extreme astigmatism. Using the telescope itself as your "glasses" will usually result in sharper images and a wider visible field of view.
Telescope is out of collimation.	Check collimation using procedures in this chapter. If optical alignment is off, perform rough and fine collimation procedures.
Telescope optics are extremely dirty. Very dirty telescope optics can result in considerable loss of contrast and sharpness.	Follow optical cleaning procedures in this chapter. Optics dirty enough to cause performance degradation will be immediately obvious to the eye.
Telescope optics are defective.	If defective optics are suspected, perform star test. If results of evaluation indicate faulty optics, CONTACT MANUFACTURER.
Eyepieces being used are poor in quality or are dirty.	Use higher-quality oculars. Except for low powers, Orthoscopic, Plössl, or more sophisticated designs are best. If eyepieces are dirty, they may be cleaned with a lens tissue and breath condensation.
Magnification being used is too high for current seeing conditions/telescope aperture.	The maximum usable magnification for a given telescope is usually said to be 60× per inch of

Table 8.1. (continued)

Symptom	Probable Cause	Corrective Action
		aperture (480× for an 8 inch telescope). But poor seeing conditions, not enough telescope cool-down time or an inexperienced observer will often limit maximum usable powers to much less.
	Optical elements (corrector plate or eyepiece) have become covered with dew.	Early dew buildup will be subtle. The first sign of this problem is that bright objects will develop halos of scattered light. As more dew collects on your corrector, detail will be lost. Use a dew shield, a hairdryer-style dew zapper gun or heating elements to prevent dew accumulation.
	You are attempting to observe through window glass. Many beginners, leery of cold winter weather, attempt to observe through a closed window. Modern plate glass is surprisingly flat, but some aberrations will still likely be introduced. Opening the window doesn't help. Warm air from inside the house will flow out through the window, creating turbulence and ruining seeing.	In normal circumstances, do not attempt to observe through an open or shut window. Take your telescope outside!
	Defective or poorly made eyepiece filter. An eyepiece filter must be optically good or it will spoil your view. You must also refocus the telescope when you add a filter of any kind to the eyepiece.	If image is sharp and clear without filter, replace defective filter. Or refocus telescope to accommodate filter.
	Defective star diagonal. A poorly constructed or defective star diagonal can affect telescope alignment and introduce image aberrations of its own. It can also affect collimation. A tip-off to star diagonal alignment problems is that the image in the field moves a substantial amount if the diagonal is rotated to a new viewing position.	Remove star diagonal, place eyepiece directly in visual back and check for image improvement. If image improves, repair or replace star diagonal.

Problem	Cause	Solution
	Tree limbs or other obstructions are blocking your telescope's view of an object. It may appear that you are in the clear, but an object like this may actually be obstructing your scope. This will appear at first as a halo, a diffuse light around the object, and will cause a loss of details. Defocusing the image by a large amount will reveal obstructing objects by their outlines against a defocused bright star or planet.	Move telescope until its view is clear of obstructions.
Moon and bright planets display blue and red coloration around their limbs.	Differential diffraction caused by thick air near the horizon will give one limb of a planet or the Moon a red colored border and the opposite limb a blue border. This is *not* caused by chromatic aberration in eyepieces or the main optics. This effect can also be caused by tube currents in an uncooled telescope.	Allow Moon or planet to rise about 30 degrees above horizon; observe while still at least 30 degrees above horizon. Allow telescope sufficient cool-down time.
Image moves in eyepiece when focus control is turned.	Caused by the moving mirror focusing system of SCTs. Small amounts of mirror shift during focusing cause the image to move. Up to about 30–45 arc seconds of shift is normal in a mass-produced SCT.	If shift is excessive, try moving mirror to both ends of its travel using the focus control. This will help redistribute grease. If shift is still excessive, CONTACT MANUFACTURER for assistance.

Telescope Mounting Problems

Problem	Cause	Solution
Telescope shakes and image vibrates in field of view in windy conditions or when telescope is touched.	Too light or improperly assembled telescope mounting. Overly windy conditions. Unbalanced OTA. Some telescope vibration is to be expected unless your SCT is on a massive permanent mount. If the image settles down within 2–3 seconds after a gentle tap on the tube, your mount is performing adequately. Almost any telescope will shake badly in windy conditions. Any vibration will be enhanced in high-power eyepieces.	Ensure that all tripod and wedge hardware – bolts, nuts, etc. – is sufficiently tight. Tripod spreader should be firm against tripod legs, but not over-tight. Learn to exercise a light touch when focusing or purchase a remote-control electric focus motor. If telescope tripod legs are equipped with both rubber crutch tips and spikes, remove tips to reveal spikes when setting scope up on soft grass or dirt areas.

Table 8.1. (continued)

Symptom	Probable Cause	Corrective Action
		Purchase shock-isolation *shake ender* pads (Celestron) for use under telescope legs.
		An unbalanced condition of the scope tube will add to vibration problems. Balance tube using optional balance weight systems if necessary. German equatorial mounted telescope can usually be adequately balanced by moving the counterweight along its shaft (an additional counterweight may have to be purchased in some cases) or the telescope along its mounting rail.
Electronic/Electrical Problems		
Telescope electronics are dead. Drive motor(s) do not run.	No power to system. Batteries low, cords and connections faulty, or fuses blown.	Check to be sure you have turned the switch to on. Listen for sound of motor or check high speed slewing to make sure you don't just have a burned out indicator lamp. Test batteries using battery tester (not just a volt/ohmmeter). Check power cords for continuity. Replace batteries and repair cords as necessary. Ensure all connections are secure. Check status of fuses inside telescope drive base or in DC line cord (refer to manual for location/presence of fuses). For troubleshooting, if scope is normally used with DC, check operation with AC if possible. If normally run on AC, check to see if telescope operates with batteries. CONTACT MANUFACTURER.

Symptom	Cause	Remedy
Abnormal electronic indications at power-up, especially in computerized telescopes.	Fuses in drive base or DC line cord blow repeatedly.	Check for possible binding by telescope motors. Disconnect any electrical accessories. Check for damaged power cords and connectors. Check to see that fuses seat securely in holder. With most of these indications, unless fix is obvious – frayed power cords, etc. – it will be necessary to CONTACT MANUFACTURER for assistance.
	Computer/electronic malfunction causes telescope hand controller display to be frozen. Or, telescope self-test (indicated by lighting a sequence of indicator lamps) does not complete correctly due to hardware/firmware problems.	Power telescope down, wait 90 seconds and restore power. Try operating from alternate power source; these types of symptoms can indicate low battery power or a defective/too long AC extension cord. If this does not clear problem, or if it recurs, CONTACT MANUFACTURER.
Drive/Tracking Problems **Telescope does not track at proper rate. Objects always drift out of eyepiece in a relatively short period of time.**	Poor polar alignment. This is definitely indicated if drift is in declination (north/south).	See instructions in this book for performing an accurate drift polar alignment. Merely pointing scope at Polaris will not provide an accurate enough polar alignment to stop all drift, but will be adequate for most visual observing.
	Right ascension lock not secure.	Check RA lock, lock if off, increase tension if partially on.
	Improper drive rate selected.	Select proper drive rate. Differences in drive rate are small, but selecting "lunar rate" for observing stars, etc., can add a small amount of drift.
	Improper adjustment/functioning of internal drive corrector on DC telescopes. Faulty motor/electrical connections on DC or AC driven scope.	CONTACT MANUFACTURER for advice/service.

Table 8.1. (continued)

Symptom	Probable Cause	Corrective Action
	Faulty external AC drive corrector on synchro motor style telescopes.	Disconnect telescope from drive corrector, plug directly into AC/mains. Repair or replace (AC) external drive corrector.
Telescope drive does not respond to button presses on hand controller.	Improper mode to allow hand controller operation, faulty hand controller.	Check manual to be sure buttons active in current telescope mode. For example, N/S buttons may be disabled during PEC training with some telescopes. Use high-power eyepiece and watch star field for appropriate movement while pushing button. 2× sidereal movement may be too slow to easily notice if a low-power eyepiece is used.
Image jitters or shakes, especially at high power, when RA drive motor is turned on.	Stepper motor drive is causing vibration as it moves in discrete steps, usually due to misadjustment/misalignment of gears and/or RA drive motor.	CONTACT MANUFACTURER FOR ASSISTANCE.
Goto Telescope Abnormalities **Computerized goto scope does not track accurately during alt-azimuth operation.**	Inaccurate or faulty initial alignment.	Turn telescope off and back on after 90 seconds. Repeat alignment process following instructions in manual. If this does not clear problem, CONTACT MANUFACTURER.
At power up, or at any point during operation, one or both of the telescope	Declination or right ascension runaway caused by electronic/ computer/electromechanical malfunction or faulty cable.	If the affected axis has an associated external cable, power down telescope, remove cable, check continuity and connectors. If problem persists,

Problem	Cause	Solution
motors begins to spontaneously operate at high speed.		CONTACT MANUFACTURER.
Objects are not in field of view following goto slew.	Poor or faulty alignment.	Power down and repeat alignment following manufacturer's instructions. Be sure to choose appropriate alignment stars per instructions. "Sync" telescope on a star in area of interest if this is an option. Use *High Precision Pointing* for maximum accuracy with Meade telescopes. If accuracy still poor, CONTACT MANUFACTURER.
	Inaccurate entry of initial site data.	Check to see that positional data (latitude/longitude), time, and other required setup entries have been entered accurately and correctly. Make sure TIME ZONE is entered correctly! Reenter as necessary.
Accessories Objects in cross-hairs of finder do not appear in main scope eyepiece.	Finder alignment faulty.	Align finder using procedures found in Chapter 7.
Declination motor does not run.	Power problems. Declination lock not secure, faulty motor.	Check power connection for integrity with push-on declination motors. For all declination motor problems ensure that declination lock is locked securely. Declination slow motion will not operate with lock unlocked. CONTACT MANUFACTURER of declination motor for advice.
Focus motor inoperative.	Faulty focus motor, bad connection, faulty hand controller.	Check connections. Ensure motor installed and adjusted per instructions. Power down scope, unplug and replug motor and hand controller.

Chapter 9

Keeping Your CAT Happy: Hints, Tips and Projects

Some of the best ideas for SCT users don't appear in manufacturers' instruction books. These are the little ideas and tips passed around at astronomy club meetings or, more recently, on the Internet in those spots where "CAT fanciers" gather. Following is a selection of tips and projects that were contributed by members of the author's Internet SCT User Group, who were kind enough to share their expert knowledge in the interest of making your SCT "better," easier to operate, and more user friendly.

Cases

This is an important topic for recent SCT purchasers. You need a sturdy and lightweight container in which to store and transport your CAT. In past years, this wasn't a concern, as a carrying case was included in the purchase of the telescope. But with the phase-out of cases by the SCT makers to keep costs down, this has become a real problem. You can still purchase a case from the manufacturers, but this can be a very expensive item. The same goes for containers supplied by photographic equipment manufacturers. These store-bought containers will do a fantastic job of protecting your scope, but a new amateur who has just purchased an expensive SCT may not be able to turn around and buy a fancy case right away. A good case is *not* an option, though, despite what the scope makers say. It is a *necessity* from the beginning. There are, however, some alternatives to paying large amounts for custom cases.

Pick-up Truck Toolboxes

In many areas, especially in the United States, it's possible to buy large toolboxes designed to fit in the backs of pickup trucks next to the cab. Some of these are expensive items, being made of aluminum. But, increasingly, plastic versions are available. These are both very sturdy as well as inexpensive. Some models even feature such niceties as locks and wheels. The only problem with these plastic boxes is they may be too large for transport in a small automobile. They can still do a fine job of protecting the telescope at home, though. Ideally, you should try to find a toolbox of the right dimensions to accommodate the foam your scope was shipped in. Remove the foam from the cardboard shipping container, place it in your pickup truck toolbox, insert CAT, and away you go. If necessary you can pad the toolbox with more foam if it is a little large. You can even use cardboard or styrofoam to take up the extra space if high-density foam isn't easily available in your area. If the box is a trifle small, you can trim your telescope's foam inserts for a good fit, but don't use a box that causes you to trim away so much foam your telescope is no longer adequately protected.

Ice Chests

An ice chest (cooler) wouldn't seem to have the makings of a telescope case, but if you can find one large enough to hold your telescope it can make a fine container; one that will be both tough and fairly compact. Check around at your local outdoor supplier and you'll no doubt run across an ice chest large enough to accommodate both your scope and its original foam. A cooler may be more expensive than some other case-substitutes, but it has significant advantages. The insulation inherent in the design of an ice chest provides a substantial amount of protection for your CAT in addition to its packing foam. A cooler will also be moisture-proof and will close securely. You could do much worse than a big ice chest.

Rubbermaid-Style Containers

America's Rubbermaid Corporation and a host of other manufacturers are now making large heavy-plastic storage containers for use in organizing and storing household items. These bins are equipped with heavy-duty snap-on lids and are available in sizes adequate for scope and packing foam. You may even locate one of these containers with a hinged lid, wheels and a handle. Because these plastic boxes have relatively thin walls, it is *vital* that you install all of your scope's packing foam to keep your CAT protected from the bumps and bruises which can come even when just transporting it to the backyard. Storage bins of this type are usually found at discount and department stores.

A Homemade Wooden Box

If you're handy with tools, a visit to the lumberyard for some inexpensive plywood may provide you with a serviceable case that's even cheaper than a plastic storage bin. While any type of wood can be expensive these days, ¼ inch plywood can still be cheap if you don't mind a knothole or two. You could resort to thicker lumber than ¼ inch, but it really isn't necessary, and using this thin stock will keep the price and weight down. The last thing you want is a case that's so heavy you'll be reluctant to use it for scope transport. All you really need is a box of some kind to *contain* your telescope's packing foam. It doesn't have to be built like a tank. The cardboard case your scope arrived in actually does a good job, but will, of course, not hold up under constant transport to dew-heavy dark sky sites.

What you want to do is build an analog of your cardboard shipping container. Measure carefully to get things just right (the old saying, "measure twice, cut once" should be kept in mind). The plywood sheets, the sides and bottom of the box, are joined using pieces of 2×2 lumber cut diagonally into triangle-shaped cross-sections. Simple drywall screws and a little wood glue will be all that's needed to hold your creation together. The addition of a hinged lid, and a couple of carrying handles complete your custom telescope box. If you wish, you can even stain and varnish the exterior. If you have some knowledge of woodworking, or can enlist the aid of a friend who does, you may wind up with a telescope case that's both attractive and functional, one that has your fellow observers constantly asking, "Where can I get one like that?"

Foam for your Case

What if you should need replacement foam for your SCT case? What if the original foam has become damaged, dirty or was accidentally discarded when the telescope arrived? The first problem is locating packing foam of any type. Check plastics manufacturers and places selling material for mailing or packing. You might even approach wholesale vendors of this type of material. When they hear your sad tale of a foamless telescope, they may be willing to sell you a small amount "just this once."

Some amateurs worry about foam outgassing. They're concerned if they don't get a foam made from just the right material or formed with just the right type of gas, it may emit vapors which will cloud or otherwise damage their optics. This *may* be a valid concern, but I've found that it's just about impossible to determine what the right type of packing foam *is*, much less locate it. I think if foam of any kind is allowed to air out until it has *no scent*, or at least just a minimal odor, it will likely be safe for use with your precious optics. Usually any odor will disappear if new foam is allowed to air-out for a few days. I've used a large variety of packing foams in the course of storing various optical components over the years and have never experienced a problem in this area.

You've managed to get some nice dense packing foam perfect for cradling your CAT. But what do you do with it now? How do you cut the stuff cleanly and evenly? A sharp, serrated carving knife will help, but it's still not easy to cut springy foam into the proper shape. There is a way. Pour water over your foam, squeeze out as much as you can, and place the entire thing in your freezer (you'll have to locate someone with a large stand alone freezer if you don't have one). Leave your foam in the cold until its good and firm, remove and cut. You'll find that what was once a daunting task is now laughably simple. Normally, foam has a tendency to tear rather than cut cleanly, even under the sharpest knife – freezing eliminates this propensity to tear. To make things really neat and easy, design a template on paper to use as a guide while cutting your foam. If you squeezed as much water as possible out of the foam after you wet it, it should not be damp at all following the cutting. It may be a good idea, though, to let the foam air out in a dry place for a few days before you use it, just to be sure.

Optical Issues

Collimation

Wrench Handles

One of the hardest things about SCT collimation is the fact that, unless you have really long arms, you may not be able to turn the collimation screws on your larger-than-5-inch CAT's secondary mirror mount without taking your eye away from the eyepiece. But adjustment is much easier if you can watch the image of a star while you're moving a collimation screw.

An easy solution is to make extension handles for your Allen wrenches. A length of wooden dowel or a piece of coat hanger wire can be taped to an Allen wrench to allow you to move the screws while looking at a star's image. Some amateurs make collimation adjustments easier still by making *three* wrench/handle combinations, inserting one in each collimation screw before beginning an adjustment session. With a wrench in place in each bolt, things go especially quickly, as there's no need to remove the wrench from one screw and fumble it into another in the dark.

Sadly, you can't use this labor-saving tip on the newest Celestron SCTs. These have Philips head screws instead of Allen bolts for their collimation adjusters, and there's no way to keep the screwdriver in place while you are at the back of the scope.

Telescope Focusing

Improving your Focus Control

The pretty chrome-plated focus knobs sported by modern SCTs look great. Their silver color stands out beautifully against the dark blue or black of the OTA. But they don't really *work* as well as they *look*. The main problem is that they're too small in diameter, making them hard to find in the dark, difficult to manipulate on cold nights by gloved fingers, and not as capable as they might be of fine adjustments. A larger-diameter focus knob would be a big relief for Meade owners, especially. The focus action in Meade SCTs tends to normally be more stiff than that in the Celestrons, making it harder to adjust the Meades in very fine increments.

Peanut Butter Jar Lid Focusers

How can you make your focus knob bigger? The possibilities are endless. Any number of everyday objects can be used. Lids from empty food jars are quite popular and are just about the right size to make a new knob for an 8 inch or larger CAT. Peanut butter jar tops are especially favored, since some brands have textured ridges around the circumferences to make it easier to grip the lid for removal from the jar or, in our case, to focus our telescopes! Very desirable are the tops from the Jif brand of peanut butter. These are exactly

Figure 9.1. Your CAT's focus control can be improved with a lid from a peanut butter jar.

the right size and are made of plastic molded in an attractive blue color. If you can't locate Jif peanut butter in your area, I would bet you could convince an Internet astronomy buddy to mail you a Jif jar top.

To make a jar-lid focus controller, locate a top that is of suitable diameter to make a good focus control knob. If you've got a small CAT, find a small jar lid. If you've got a C14, you can make your add-on knob as large as you want by using a top from an institutional size food product jar. Cut or punch a hole in your jar lid that is just slightly larger in diameter than your focus control. Take care to center this hole as accurately as possible in the lid. Wrap a few layers of tape around the focus knob, and slip the jar lid over it, adding tape if necessary to ensure a snug, nonslipping fit. For a more professional job, you visit your local auto parts store and locate a rubber grommet of just the right size to slip over your scope's focuser. Measure your control's diameter beforehand, and choose a grommet slightly smaller in diameter than this. This may be the best solution of all – it's the one I used, anyway. Drill or cut the hole in your jar lid to accommodate your grommet, glue the grommet in place with contact cement or Super Glue, slip your completed "focusing circle" over the scopes original knob, and enjoy!

Owners of the smaller CATs, C5s for example, may find that the big lid from a food jar is just too large to use, and may not even fit due to the smaller size of the telescope rear cell. A larger focus control may be even more helpful for these telescopes than their bigger brothers, however, since the knobs on many small SCTs and MCTs are way too small for comfortable use. A babyfood jar top is one possibility.

Dealing with Focus Shift

Focus shift, the movement of an image in the field of an SCT as you turn the focus control, is the bane of SCT users. It can be very irritating, even if it doesn't affect focusing or viewing. At really high power it can actually move the object of interest right out of the field of view. Focus shift is inherent in the SCT design because of the moving mirror focusing system. But it can be made worse when the grease on the baffle tube the mirror slides on becomes unevenly distributed in the course of normal use. Often, moving the mirror through its entire focus travel a couple of times will serve to redistribute the lubricant and will really reduce focus shift. It might even be a good idea to do this on a regular basis. Perhaps every time you check your scope's collimation.

Focusing the Telescope: Hartman Masks

You've built and installed a peanut butter lid focusing knob and you've eliminated all the focus shift you can. But you still find it difficult to focus your telescope accurately, especially when looking at the dim image in your 35 millimeter camera's viewfinder when you're doing astrophotography. Focusing star images on a computer monitor when you're doing CCD imaging isn't as easy as you thought it would be either. Isn't there some way of obtaining exact focus other than by just trying to judge an object's size and sharpness? There certainly is. You can devise a simple tool called a *Harman focusing mask* to make focusing much easier.

A Hartman mask is simplicity itself to make. Obtain a piece of heavy cardboard or poster paper (black is nice). Cut out a circle of the material just a little larger than the end diameter of your telescope's tube. You'll then cut two small holes into this circle. These should be directly opposite each other and should be positioned near the edge of the circle so that when it's placed over

Figure 9.2. A Hartman mask is easily made and can make focusing incredibly easy.

the end of your scope, the holes will be right on the edge of your corrector plate. The size of these small holes is not overly critical, but about one fifth of the diameter of your primary mirror works well. That would make the holes about 1.5 inches in diameter for an 8 inch telescope.

Use a compass to draw the holes and cut them out with a sharp blade. An Exacto knife works perfectly. You'll place your mask over the end of your scope, so you'll need to devise a way to hold it in place. Gluing/ taping a cardboard strip around the circumference to act as a lip will work great. If this seems to be too much trouble, you may merely tape the mask over the end of the scope with a few strips of masking tape. Remove your dew shield to install the mask, fit it into place over the corrector and it's ready for use (the mask must be close to the surface of the corrector plate for this little trick to work, so you must do without your dew shield while focusing).

But what does a Hartman mask do, and how can it help your focusing? Point your scope at a bright star with the mask in place and look thorough an eyepiece. Due to the twin apertures, you'll see not one star but two. They'll be separated by a small distance and will look just like a binary star. Turn your focus control experimentally. If the stars move farther apart as you turn, or just start to blob out, turn the control the other way. You'll see that as you turn the focus knob in the correct direction the two star images move closer and closer together and finally merge. When you've got the images perfectly merged (if you move past focus, they'll separate again) remove the mask and take a look at the star. You'll find that, almost magically, you've achieved perfect focus! It is incredibly easier to focus by merging two stars than it is to judge when a star is sharpest by observing its size and brightness.

Bungie Cord Focusing

You don't really have to use genuine elastic bungie cords to use this focusing trick. Any type of line or rope about ¼ inch in diameter will work well. Tape two lengths of cord in front of your corrector plate. Remove the dew shield, and arrange the cords so they form a cross-hair at the front of your SCT. Be careful, of course, not to allow them to touch the corrector plate. Think Newtonian telescope secondary mirror support (spider). What we want to do is to add a cross-shaped diffraction pattern to stars, just as is seen in a Newtonian reflector. Observe a star with your cords in place, and you'll see it has sprouted diffraction bars. Focus until these diffraction spikes are sharpest, and you'll be in good focus. This method reputedly works especially well with CCD cameras.

Dew Problems

A Homemade Dew Shield

The first line of defense against dew for the SCT or MCT user is a dew shield. A metal or plastic extension for the end of the tube provides some protection for the dew-attracting corrector plate. You simply must have one if you live anywhere with humidity high enough to cause much dew formation. The only problem with dew shields is the *cost*. You wouldn't think a simple plastic tube could be so expensive! But they are. Luckily, a dew shield is such an easy thing to make there's really no excuse for buying one rather than making one of your own.

The most convenient style of dew cap is one that's normally a flat piece of heavy duty plastic and which is formed into a tube only when it's placed over the end of your telescope. A dew shield that can be stored flat takes up much less room at home, and, especially, in your over-laden car when you're transporting your telescope and all its accessories to a dark site. These shields are typically made from sheets of heavy-duty plastic and are fastened into a tube with Velcro. The hardest part of making one of these is finding a suitable sheet of plastic.

The most perfect material is a plastic sheeting used for wall covering whose brand name is Kydex. This material can be obtained in a range of colors, with the flat black desired by dew shield makers being easily obtainable. Easily obtainable if you can find any Kydex at all, that is. Your best bet is a local plastics distributor. If you can't obtain Kydex, a trip to the hardware or home improvement store will offer some alternatives. Plastic sheeting designed to be used to help in bagging leaves is a fairly common item in lawn and garden emporiums. This is material used to help keep a trash bag open when you're dumping leaves into it. But we don't care about that, what matters is that it's a plastic sheet the right size and thickness to be formed into a dew shield.

Before purchasing a sheet of dew shield material, you should ascertain how big a piece you'll need. It should be long enough to go around the circumference of your tube end with about 2 inches extra of overlap for fastening. The other dimension is determined by the size of your primary mirror. The shield should extend at least the width of your primary mirror in front of the corrector for adequate protection. A few inches more than this is even better.

After you've bought the dew shield material, stop off at a fabric store or department store on the way home. Hunt around until you find the Velcro fasteners. These will be used to hold your shield in the form of a tube. Purchase strips if possible, but small squares can also be used if long strips are not to be had. Obtain Velcro with peel-off self-adhesive backs if you can.

Back home, carefully cut your sheet of plastic or other material into the dimensions you determined are appropriate for your telescope. It might not hurt to measure things one more time. If you have to head down to the hardware store for more sheeting (especially the somewhat pricey Kydex) you'll soon find the money savings from "rolling your own" evaporating! Set up your SCT on its wedge inside the house and remove its aperture cover. Wrap the material around the end of your tube experimentally to make sure it fits and fasten it in place with a few pieces of masking tape. You want the fit to be snug enough that the shield does not droop. It should fit over the entire tube end-ring for best results. When you form the material into the tube, make the sky end slightly – just a small amount – smaller than the scope end for best performance. Just a little, though. You don't want to introduce vignetting. You may want to try the scope on a terrestrial scene with and without your shield to make sure you're not cutting off the view. When the shield is securely taped into form, slide it off the end of your telescope. Take a soft lead pencil and mark along the seam at the overlap point. This will provide you with a reference point for the positioning of your Velcro fasteners.

Remove the tape, and place your Velcro pads or strips under the overlap point for the dew shield. One piece will face up, mounting on the top of the edge of the shield that is overlapped by the other edge, and one piece will face down, being mounted on the underside of the edge that's on top. Allow your self-adhesive Velcro to set for several hours before trying your shield. I've found that the sticky glue used on self-adhering Velcro may not hold well unless it's allowed to set for a little while. It is also a good idea to clean surfaces with alcohol before applying the Velcro to ensure a good bond. Once the pads are secure you may wrap your new dew cap around the end of your scope, fastening it into place with the Velcro. Looks great, works great!

Aluminum Foil Insulation

Many amateurs rely on dew zapper guns/hairdryers to keep dew off the corrector. These can work well, but, if possible, you'll want to minimize their use. It gets tiresome having to run around to the front of your scope with a blow dryer every few minutes when you're hunting faint fuzzies. You can install a heating element on your corrector plate, but these cost money and result in a tangle of wires and control boxes. One solution to this problem is to improve the performance of your dew shield. Is there a simple way to do this? Yes! Wrap it – or even the entire OTA – in aluminum foil!

This treatment may make your scope look a little funny, but many observers swear this insulating layer can double or even triple the effectiveness of the dew shield. You can use plain, old household aluminum foil out of the kitchen. This will work well if all you want to do is cover the dew shield. But if you want to insulate the entire scope, a cover made from an aluminized Mylar space blanket will be even better. You can make a little snug-fitting sweater for your CAT from one of these blankets. Hold in place with tape. With a little ingenuity, you can rig your CAT a covering that can be removed and reused many times.

Speeding Cool-Down Time

I happen to live in the sunny southern part of the United States where a winter's night with a temperature below 32 degrees Fahrenheit is a rare one. Most evenings, even in November and December, are more likely in the 40s or even 50s. But I still find that I have to allow some time for my CAT to adjust to outside temperature after being taken from a warm house (or to equalize with the outdoor temperature in summer when taken from an air-conditioned house to a warm, muggy backyard). On a weekend, this is not a problem. But this waiting period becomes a pain when I'm engaged in a spur of the moment weeknight observing run. I shudder to think how long it must take for SCT optics to adjust on really cold northern nights! You could store your scope permanently in an unheated (or uncooled) shed or garage. But these locations are often not the best, either for cleanliness or security reasons. Ah, but there's a way to speed the cool-down process!

Many owners of large Dobsonian telescopes have found that small battery-operated fans located at the mirror end of their telescopes' tubes can greatly decrease the time it takes for the mirror to cool to ambient temperature. It is only now that CAT owners are catching on to the fact they can do the same thing.

Procure a little CPU fan or CPU cooler. These 12 volt devices are designed to sit atop your computer's central processor chip and provide a cooling flow of air from a tiny, high speed fan. You'll find one at your neighborhood computer supply house for very little money.

When you've found a suitable fan, procure a tube of some kind that can fit either *over* or *into* your telescope's visual back. A piece of sink drain tube (PVC) can be obtained which will fit nicely into place, being the same diameter as an American Standard eyepiece. If you can't find a suitable drain pipe, the cardboard tube from a roll of paper towels or toilet paper will fit over the visual back if you remove the set-screw. If necessary, you can place some tape around the inside of your tube's end to make the fit more snug. The length of your tube is not critical.

You can now affix your little CPU fan to one end of the tube. Arrange the fan so that air blows *into* the telescope. How you attach it will depend on the size and shape of your fan. It may be possible to glue it to PVC pipe with contact cement or Super Glue. If you've opted for paper or cardboard tube, you can simply tape the fan in place – air-conditioning duct tape works well for this. Seal any gaps between fan and tube with more duct tape.

You also have to rig a means of connecting your fan to 12 volts DC. The simplest way to provide power is to purchase a male cigarette lighter plug and cable at the same time you buy the fan. Connect the leads to the leads coming off your fan (observing the proper + or –

polarity, which should be indicated on the fan's wires or packaging). Solder these into place, wrap some electrical tape around the connections, and you're ready.

Slip your fan's tube assembly into or over the visual back, plug in the power, and ambient temperature air will be blown into your OTA making cool-down go really fast. There are enough gaps in your OTA – around the corrector especially – to allow sufficient air-flow through the tube. Some people worry about blowing dust into the tube, but this has not been a problem for me. If this worries you, a small pad of filter material – perhaps from an air-conditioning system filter – positioned over your fan's intake should help keep that bad old dirt out of your lovely scope.

Telescope Mounts and Tripods

Tripod and Mounting Stability Issues

Is it just my imagination, or have telescope tripods become steadily flimsier over the years? Many CAT tripods could stand to be a little stronger than they are. The smaller GEM mounted telescopes, in particular, can be problematical. In a quest to keep prices down, the makers of the import mounts these telescopes ride on have deserted wooden tripods for flimsy aluminum affairs. The actual equatorial heads included with these telescopes are often quite good – surprisingly good – but they are prevented from living up to their potential by shaky aluminum contraptions that don't deserve to be called tripods. If you've taken care to tighten all of your tripod's bolts and nuts, but still find your scope much too shaky, there are some things you can do to steady that sucker down.

Milk Jug Weights

A quick cure for a shaky tripod is to suspend some kind of a weight from the underside. A 1 gallon plastic jug or bottle of some kind filled with water is perfect for this purpose. One gallon milk jugs work great as they have integral handles molded into them that allow the jug to be easily suspended from the scope's tripod. Obtain some strong nylon line and tie one end securely around the top of the tripod so it extends down between the legs. Knot the free end to the handle of the water-filled jug so it is suspended a couple of inches from the ground. You should now find your telescopic views considerably steadier. This scheme works especially well at holding the scope steady during focusing. Be careful not to bump the jug while you're observing, though, as moving it will introduce swaying that will ruin your views and which may tend to make you feel seasick!

An Accessory Tray/Leg Spreader

One reason for the unsteadiness of the tripods of entry-level fork-mount telescopes is that many lack leg spreaders to hold the legs of the tripod securely apart. But one can be made quite easily. Start by setting up your tripod and ensuring its legs are as far apart as they will go. Measure the distance between the legs about halfway down the tripod. Then lay out a triangle on a piece of ¼ inch plywood or other sturdy material, with each side being the length you just measured. This will become your spreader. Head down to your local auto parts store and locate three hose clamps like those used to secure automobile radiator hose ends. Make sure they are big enough to fit over the legs of your tripod. You'll now affix each hose clamp to a triangle vertex with wood screws that extend through holes you drill though the clamp bands. Slide the three hose clamps, now attached to your plywood triangle, over the tripod ends and up until the tripod legs are pushed firmly apart. At this point tighten all three hose clamps into place. Your legs are now held apart, making for a much steadier scope. You may bend the hose clamps until they are at an angle matching that of the tripod legs, resulting in a firm and attractive fit.

If you wish, you can turn this spreader into an accessory tray. Nail and screw small strips of wooden molding along the edges of the triangle to provide it with a lip to keep small items from rolling off. You can also drill holes in the plywood in 1.25 inch and 2 inch sizes to hold your eyepieces (a hole saw can be obtained at the hardware store for very little money). Finish by sanding, staining and varnishing your new tray. Or you can just slap on a coat of black exterior-type paint. Whether your spreader/tray looks proletarian or professional, it will go a long way to ending telescope vibration.

Felt Pads as Shock Absorbers

You may find that your new German equatorial is just too prone to vibration. A proven cure for this is to place a layer of felt fabric between the tripod head and the equatorial head. Usually, the mount-head part of a GEM can be removed from the tripod, often by undoing a single bolt. Separate your tripod and mount and measure the top diameter of your tripod. Cut a circle of felt just the right size to cover the top with no overlap. Cut holes in the fabric appropriate for the bolt or bolts that fasten tripod to mount. Replace the GEM head and tighten everything down. You should find vibration now to be considerably reduced. You would *think* that a material with more shock absorbing qualities, rubber sheeting perhaps, would work even better than felt in this application. Alas, this does not seem to be the case. Rubber does not seem to damp out vibrations nearly as well as felt when used between tripod and mounting.

Take Those Crutch Tips Off

The leg ends of many SCT tripods are covered with rubber crutch-tip-style ends. Many SCT users just assume this is the way the tripod is and never think to remove these ends. Rubber tips can work well when a telescope is place on a hard surface like an asphalt or concrete parking lot. They work much *less* well when the scope is set up on a soft dirt or grassy area, however. Far from absorbing vibrations, these tips actually may *add* to the scope's shakiness when it's stationed on soft surfaces. The solution is to take the tips off. Some scope makers have shown foresight by equipping the tripod legs with spike-type ends that are revealed when the rubber tips are removed. If you don't find spikes after you take off your rubber tips, you can use your scope as is, open ends and all. You'll probably have to clean dirt and grass out of the ends when the observing session is over, but that's a small price to pay for steady viewing. You may be able to plug the ends if they're open, perhaps by using wooden dowels glued into place with contact cement or silicone adhesive.

Ersatz Shake Enders

Celestron sells an accessory called Shake Ender Pads. These tripod footpads are shock-absorbing disks that fit under each tripod leg to eliminate vibration. They are a very desirable item and really do reduce telescope shakiness. They are also relatively *expensive*. This seems surprising at first, but close examination reveals they are not as simple as they look. They look similar to the pads that go under furniture legs to prevent scuffing, but they are actually cleverly made, consisting of an inner cup that the tripod leg end rests on which is suspended from the rest of the shake ender by a layer of special "sorbothane" rubber. Could the enterprising amateur make her or his own? Certainly! Homemade shake-suppression pads may not be as effective as the real thing, but when combined with other vibration reducing strategies, they can help end your shakiness problems.

The simplest homemade pads are nothing more than upside down bathtub drain stoppers. The best stoppers to use for this purpose are those with an inner ring molded into their surface. This ring would normally face down, and would be inserted into the tub drain, but we'll place the tripod leg end into this depression. If the opposite side of the stopper has a narrow raised portion for attachment of a metal ring and chain, this may be trimmed off with a sharp hobby knife. If there's a wide raised area, just leave it in place, as it will add to the stopper's vibration-reducing characteristics. Remove the metal pull chain, if present, of course. Placing three tub drain stoppers under your tripod legs might seem to be a slightly humorous and overly hopeful method of trying to stop your CAT from shaking, but users report that this really works. The person who suggested this trick to me experienced a reduction in "shake time" from 5 seconds to an excellent time of less than one second!

What else can be pressed into service as homemade Shake Enders? Well, what *looks* like a Shake Ender? The aforementioned furniture carpet or scuff protectors for furniture. Sadly though, these, as you'll quickly see if you try them, do little or nothing to reduce the shakes. They can actually *add* to scope vibration, believe it or not. But this can be fixed! Obtain three carpet protectors, with those made from hard rubber being the best choice. Then stop off at the office or computer supply store on the way home and buy a couple of neoprene mouse pads. At home, take your three carpet protectors and your mouse pads and make little sandwiches. Cut neoprene circles from the pads the same diameter as the carpet protectors. Place one circular cutout piece of mouse pad on the bottom of each protector and another piece on top, where your tripod leg end will rest. The pieces of neoprene can be glued into place with contact cement. Place a completed "shake ender" under each tripod leg and enjoy.

Fill Tripod Legs With Sand

You've tried all the above methods of eliminating scope vibration, but there is still too much shaking for your taste. As a last resort, you can try a trick that many scope users, especially those saddled with the modern light extruded aluminum tripods, swear by. Fill the legs of your tripod with sand. This can, in some instances, have a dramatic effect on scope steadiness. You'll want to use clean sand, but this is easy enough to find at a lumberyard or home improvement store, as it is an integral part of concrete mixtures. One bag should be far more than enough. How you fill and seal your tripod will depend on the particular model you have. Often, the easiest way will be to pour in sand from bottom of the legs after having removed their tips. Before you do this, remove the telescope and equatorial head, of course. You may find it necessary to seal some small gaps in the tripod to prevent sand from leaking out. Any type of silicone sealer will work well for this purpose. Make sure that you can remove the sand if you don't notice any improvement – don't glue tip ends and other openings permanently.

Wedges and Polar Alignment

Smoother Wedge Azimuth Motion

Most SCT wedges have pretty smooth altitude motions, especially those that have been equipped with deluxe latitude adjusters. But just about every wedge/tripod combo I've seen falls down when it comes to side-to-side movement. But easy azimuth motion is just as necessary when you're trying to do an accurate polar alignment as smooth altitude movement is. What can be done?

The most creative solution I've seen involves adding a slick bearing surface to the underside of your wedge. What you're trying to do, in effect, is add a smooth Dobsonian telescope-style bearing to your SCT's equatorial wedge. The perfect candidate is an unwanted CD. Not only is the surface of a compact disk smooth, there's even a predrilled hole, perfect for Meade telescope users. If you're using a Meade wedge, just remove the wedge from the tripod, slip the CD over the central bolt (you probably won't even have to enlarge the hole in the disk) and replace the wedge. You'll simply be amazed at how much easier your wedge now moves. Celestron scope users may find this idea a little harder to implement since the Celestron wedges are affixed to the tripod, not by a central bolt, but by three radially positioned bolts. You'll have to drill three holes or slots in the CD to accommodate this layout.

"Permanent" Polar Alignment

You've spent a lot of time accurately polar aligning your telescope in the backyard. Is there some way to avoid having to do this every time you need a dead-on polar alignment? It *is* possible to preserve your alignment by marking the exact positions of the telescope's tripod legs. If you need only middling good accuracy, you can do this by placing some simple marks on the ground. These can be three stakes driven into the lawn just outside each tripod end. You can even settle for three rocks or bricks as markers. But the stakes are likely to be run over by the mower the next time you cut your grass. And the bricks won't last long under the tender mercies of neighborhood kids or the lawn care man. A better solution can be found in a few lengths of PVC pipe.

Go to a plumbing supply store and purchase a cast-off length of 3 or 4 inch diameter PVC pipe a couple of feet long. Returning home, cut-off three 6 inch long sections. Drive the ends of these into the ground at properly spaced intervals so each leg of your tripod rests in the exact center of a PVC "circle." If necessary, refill the center with dirt so it's flush with the ground. Pack this dirt down so the tripod leg has a firm surface to rest on. You only want to leave a *small* amount of PVC pipe visible above ground – ¼ to ½ inch. You don't want the pipe to *touch* the tripod legs – this could cause vibration – you only want them to act as accurate position *markers*.

When your pipes are in place and the tripod is properly positioned, do a good polar alignment. If you can see Polaris, of course, use it. If your scope has a polar alignment finder of any kind, use that too. Then follow up with a drift alignment (see Chapter 10). When you're done, make sure the wedge altitude and azimuth axes are *tightly* bolted down. Observe as you always do, and return the scope to the house when you're done. The fun part comes on the next evening! Instead of polar aligning all you have to do is set up the scope so the legs of the tripod are positioned in the centers of the circles made by the lengths of pipe you drove into the ground. Position each tripod leg end just as you did before, and you will find that your alignment is "still there." The *most* you'll have to do is do a little fine-tuning with the drift alignment procedure if you're going to be taking pictures. You should be close to the pole already and drifting won't take long at all. Naturally, if you change the altitude or azimuth of the wedge or GEM head, either purposefully to observe from another site or accidentally, you'll have to redo your polar alignment.

If driving PVC pipe sections lengthwise into the ground is not suitable for your location, there are other ways to "mark" your polar alignment. If you normally set your telescope up on a concrete pad, driveway or patio, you can make marks with paint to indicate your tripod leg positions. A nice solution, and one that's permanent, would involve pouring three round concrete pads out in the yard for your scope legs to rest on.

This is not as hard as you may think, since easy-to-use bags of concrete are available in small quantities (Sackcrete) at home improvement stores for use by homeowners.

If pouring concrete pads on your beautiful lawn doesn't appeal, building supply houses can furnish concrete paving blocks or "stepping stones" which are perfect for scope leg pads. These can be found in sizes as small as 6 inch by 6 inch and 1 inch thick – just the right size. They can even be found in various pastel colors!

If you can mark the position of your scope tripod following a polar alignment, you'll find that your subsequent observing sessions are a lot more fun. You won't hesitate to do high power planetary observing or deep sky long-exposure photography if you know you won't have to fool with exacting polar alignment every time. Marking your scope's aligned position is almost as good as having a permanent backyard observatory.

Miscellaneous Projects and Tips

Telescope Warning Lights

It's amazing how dark it can get on a moonless night away from city lights. It'll get so dark that even when your eyes are fully dark-adapted the scopes on the field will be nothing but vague shapes in the darkness. A modern SCT with a dark blue or black tube and tripod becomes almost invisible. Most of the time, this is not a real problem. You know where your scope is, and won't bump it. But occasionally your club star party may be host to some visitors or novices without scopes of their own, who tend to wander from telescope to telescope in the dark: "Hi! Mind if I look through" – BUMP! – "Ooops, sorry, was that your telescope?"

The solution to this problem is a few small red LEDs strategically positioned on your tripod. An electronics supplier can provide you with three light emitting diodes, a small double-A battery case for each, and the resistors required to drop the battery voltage enough to prevent the LEDs from frying. The package the LED comes with will often outline how to wire up the LED and what value of resistor to use. Choose standard brightness LEDs rather than the also available super bright models. You don't want to interfere with anyone's night vision.

Wire your three resistors and LEDs to the proper terminals of each battery case, tape leads and LEDs to the side of the battery cases, affix a piece of Velcro to each tripod leg and battery case and insert batteries (no need for a switch, if you don't want to fool with one, just remove the batteries to turn off your LED). Velcro each light assembly to your tripod at a height off the ground of your choice, and you'll have a nice warning light to mark the location of your scope.

If you want a *really* effective indicator – or at least one that looks cool – your electronics retailer may be able to provide you with a blinking LED assembly. If you're knowledgeable about electronics, you may even be able to devise a blinker circuit for your LEDs without trouble.

Accessory Cases

Astronomy-related accessories, especially eyepieces and other optical paraphernalia, tend to multiply at an alarming rate once you've been into astronomy for a few years. You'll begin to wonder if those 25 millimeter eyepieces are reproducing in the back of a drawer somewhere! In order to provide a safe and secure place for these items, you'll need some cases. This will become especially important once you move beyond simple designs of eyepieces and start investing in Naglers, Panoptics and other expensive designs. An eyepiece that costs as much as a good camera lens needs a more secure storage place than the cardboard box it came in.

Tool Attachés

Aluminum cases filled with cubed foam are the ultimate accessory case. They are very sturdy and can be customized to hold everything from large 2 inch eyepieces to off-axis guiders to reducer–correctors. Formerly, these cases were very expensive items; usually they were sold for carrying pro-level photo-graphic equipment. But then a wonderful thing happened. The same style aluminum attaché cases began to be produced in China. Shortly thereafter, various astronomy vendors began selling these rein- forced cases expressly for use as eyepiece cases. And they sold them for a fraction of the cost of similar photo cases. It's true that these Chinese-made wonders are not as strongly built as a "real" aluminum camera gear case, but they are more than adequate for most amateur astronomy applications.

Just when I thought the case situation couldn't get any better, *it did.* The Chinese factories started cranking out these attaché cases with abandon. It wasn't long before they started appearing in discount houses and home improvement centers for less than half what the astronomy merchants had been selling them for. Look in the tool/hardware section rather than the photo department, as these are usually advertised as toolboxes or attaché style tool cases. By "half what the astronomy houses had been selling them for," I mean *less than $20 apiece!* I've paid more than this for tiny and flimsy plastic eyepiece cases.

Other Cases

The tool attachés are really the best choice for eyepiece and equipment cases, but what if you can't locate any in your area? Or maybe you just want something a little smaller? A trip to the local sporting goods store may provide a solution. Depending on firearms popularity and laws in your area, you may be able to find nice plastic cases being sold for carrying pistols. The foam included with these units is often textured with "hills and valleys," but it can usually be flipped over to provide a smooth foam surface. The foam in these types of cases is usually not cut into cubes, so you should use the "freeze and cut with a sharp knife" method described earlier for use on foam for scope boxes.

In truth, there's no shortage of small plastic boxes available at tool and department stores which could be used for CAT equipment. The usual drawback is that there is no foam. But if you've got a source of dense foam of some kind, it's easy to make a great eyepiece box. A Rubbermaid storage container or a plastic toolbox or even a fishing tackle box can easily be converted into a great looking and great working case.

Astronomy Equipment from the Music Store

The local music equipment supplier can be a good source of astronomical equipment. First on the list of nice items for CAT owners to be found here is a *drummer's throne*. This is a small adjustable stool designed for use by drummers. Most have a range of heights perfect for use with a fork-mounted SCT. These stools are very lightly made, but capable of supporting considerable weight. They are also collapsible, making them easy to pack for a jaunt to the dark site. The only problem is that for some reason these drummer's seats seem to have been going up in price exponentially of late. Check on their prices at your music store and bounce this against the cost of the various chairs and stools sold by astronomy merchants expressly for use while observing before forking over your hard-earned money.

Another useful item to be found hidden amongst the guitars and sheet music is an *amp stand*, a folding X-shaped framework made of chrome-plated tubing that is designed to support the weight of an amplifier control head. It looks like the lighter stands you'll see in restaurants being used to support the waiter's tray at the side of your table. An amp stand is perfect for supporting a plywood sheet for use as an observing table, and folds up nicely for transport.

Musicians and astronomers actually have a lot in common. We both often take delicate equipment to distant sites. Look at the music store's case selection, paying special attention to those designed to hold amplifiers, microphones and other noninstrument equipment. Chances are you'll find one just perfect for astronomy equipment too. The only problem is the price. Musicians must not be quite as stingy as amateur astronomers!

Velcro is the Amateur's Best Friend

This material has a multitude of uses around the telescope. It can be used for everything from holding counterweights in place to providing a means to hang your telescope hand controller on the tripod. Using

Velcro keeps things organized and allows you to quickly put your hand on a controller hand box or small flashlight, since you'll always know where these items are. As to where you place your Velcro pads on the telescope, that is purely a matter of choice and functionality. For hand controllers and similar items, a good out of the way location is on the tripod's leg spreader. As noted earlier, Velcro strips and squares are commonly available both in hardware and fabric stores.

A Paint Can Lid Accessory Tray

An accessory tray mounted on your telescope's tripod is a handy thing to have. It provides a place to keep eyepieces and other small items handy for immediate use at the telescope. Unfortunately, the tripods of most fork-mounted SCTs and other CATs do not feature these trays. But you can make a nice one for very little money.

All you need for this project is the lid from a plastic 5 gallon paint bucket. Your local home improvement or paint store will offer empty 5 gallon paint cans for sale (or they may even give you one for free if you ask). All you need is the lid – you can discard the bucket or use it for another project. Drill a hole in the center of the lid the same diameter as the threaded rod your tripod's leg spreader is attached to. Loosen the knob holding your spreader in place, and remove the leg spreader temporarily. Slide the paint lid onto the threaded rod, orienting it so that the lid's lip is facing up. This will keep eyepieces from sliding off. Replace the spreader on the rod and tighten it down. The paint lid will now be resting on your spreader, providing a nice accessory tray for your use as in Figure 9.3.

Drives and Electricity

Too Much Periodic Error?

Periodic error is the small back-and-forth east/west drift of your CAT's drive due to small imperfections in the drive gearing. It is why we must guide our telescopes during long-exposure photography. If it weren't for this drift, we could open the camera shutter and walk away. It may be possible to minimize the

Figure 9.3. The lid from a 5 gallon paint bucket makes a great accessory tray.

periodic error in a new telescope. Often the small gear imperfections will be smoothed out over time as your drive operates. If a new SCT seems to have a disturbingly large amount of error, try running the drive for a few hours. You may be pleasantly surprised at how the drive "steadies down" after a while.

Power for your CAT

A Battery Box

A 12 volt lawn and tractor or motorcycle battery can easily provide all the power you need, even for computerized and power-hungry scopes like the Meade LX-200. However, using a bare battery may not be your cup of tea. A battery connected to your scope and accessories via a pair of large alligator clips is just asking to be knocked over in the dark, possibly shorting out your telescope and causing damage. If the battery is a non-sealed type (not recommended) you may find battery acid spilling everywhere. The solution is simple. Your local auto parts store can furnish plastic battery boxes (see Figure 9.4). These cost only a few dollars and are big enough to accommodate even larger deep cycle marine cells. This large size can work very well for the

Figure 9.4. An inexpensive plastic battery box provides a neat, secure home for a motorcylce or lawn tractor battery.

amateur using a smaller motorcycle battery, as it allows enough room for some wiring.

While at the auto parts store, also purchase two battery cables with terminal connectors appropriate for your chosen battery. Get the shortest ones you can find, without regard to whether they are black or red. Purchase a female cigarette style receptacle, one that attaches to a battery via two large alligator clips. Locate your battery in its box. You'll usually find that a strap or bracket of some kind is included to hold the battery firmly in place. You can then attach the cables to your battery and devise a little terminal strip for the other ends.

This terminal strip need be no more complicated than a short strip of wood. Drill holes in the strip large enough for two ¼ inch bolts. Insert these bolts through the brackets on the free ends of your battery cables, and through the wood strip, securing them with nuts and washers. Mark each bolt in some way as "+" or "–". Remove the alligator clips from the ends of the wire going to your cigarette lighter receptacle and strip the insulation off about 1 inch from the ends of the cables. Attach the bare wires to the bolts on your terminal strip observing correct polarity (disconnect the battery cables from your battery while doing this wiring). All you should have to do is wrap the bare wire around the nut end of the bolts between nut and washer and tighten things down. You may route the cigarette lighter cable out of the box through a small hole or, if you want to be fancy, you can mount the female plug receptacle on the side of your battery box.

Maintaining your Scope Battery

Inexpensive 12 volt lead acid batteries can work fantastically well as power sources, but to keep them going, they'll need a little tender loving care. Want to destroy your lead acid battery? Discharge it completely a few times. Letting a lead acid battery completely discharge often will ensure that it's soon ready for the scrap heap. Avoiding this is simple. After each use, place the battery on trickle charge (usually overnight is good – you will not usually damage a lead acid battery by leaving it on trickle charge for as long as you want). How can you tell if your battery is charged sufficiently? The only way to be sure is to check it under load. Unless it is really, really low in voltage, a voltmeter across the battery terminals will not give a reliable indication.

If you have a large battery charger, it will have a meter indicating charge status. But if you don't have one of these big, high-capacity battery chargers, there is really not much reason to buy one. Assuming your battery's capacity is fairly high, charging for a few hours with a trickle charger, even after a heavy night of use, is almost always sufficient. And trickle chargers are very cheap. Unfortunately, most don't offer any kind of meter or other indication to tell you what the battery's charge condition is. But you can purchase a small and inexpensive battery tester at your local auto accessories emporium that will be more than good enough. Look for a tester that plugs into a car's cigarette lighter. These are designed to live in an automobile's glovebox and allow stranded motorists to check the status of the vehicle's battery without opening the hood. These small units fit perfectly in your observing accessory case. They indicate charge status with a series of LED indicators. Plug this into your battery's cigarette lighter receptacle and you instantly know if the charging is complete. Some of these little testers even feature built-in lights that can easily be painted red!

Lead Acid Battery Alternatives

If you have a few dollars burning a hole in your pocket and would like a setup that's a bit neater than a motorcycle battery in a cheap plastic box, go to a local discount, camping or auto parts store. Look for a *portable power pack*. These are sealed batteries of some kind (usually gel cells) in neat plastic boxes with one or more cigarette lighter outlets mounted on a front panel. A convenient carrying handle is built in, and the power pack often has a number of other niceties. Often some kind of a light is built into the box. A charge indicator is also commonly present (the charger is usually built into the enclosure). The units found amongst the auto parts will also include a pair of jumper cables for starting your vehicle (this can come in handy if you wind up with a dead battery at your dark observing site). The most expensive units may even possess a self-contained power inverter. This will provide a source of AC for use by AC telescope drives or even laptop computers. Before purchasing a power pack, make sure its stated power capacity (in amp-hours) is sufficient for your needs.

Imaging and Photography

Balance, Balance, Balance!

The main requirement for successful long-exposure astrophotography is that your scope have an accurate clock drive. To some extent this is dependent on the quality of the motor and gear set furnished with your telescope. But you can easily increase the quality of your scope's tracking by ensuring the OTA is well balanced. An off-balance tube means that your motor and gears must strain to keep the scope moving, and any problems inherent in your system will be exaggerated.

To balance your GEM scope you just slide the counterweights on their shaft and the scope in its mounting saddle. If you've got a fork-mounted CAT, you'll need to festoon its tube with some counter-weights. This can be as elegant as a weight that slides on a dovetail bar (purchased from any SCT vendor) or as simple as a beanbag filled with lead shot and attached to the tube in various places with Velcro. The important thing is balance in RA; declination is less critical. When you're doing astrophotography, check the tube for balance whenever you move to a different part of the sky, moving the weights as required. Balance things so the SCT actually tends to be *off-balance to the east* by a *small* amount. Doing this will cause the drive to pull the scope along, eliminating any harmful slack in your gears.

Dim that Guiding Reticle

Let's admit it, it is very difficult to find guide stars using the off-axis guiding devices that most SCT-equipped photographers use. One way to maximize the number of candidate stars you have available is to keep your reticle brightness low. You want it to be *just* bright enough so you can make out the cross-hair lines. Any brighter than this and you'll blot out dim stars in the field. If the reticle can't be dimmed enough to suit you, stuff a bit of toilet tissue in ahead of the illuminator, or paint a portion of its LED with fingernail polish.

Guide while Seated

It's *nice* to observe while seated but it is, in my opinion, *necessary* to guide deep sky photos while sitting. Trying to stare at a dim star for a half-hour or more without ever taking your eye from the eyepiece is a difficult task even if you're comfortable. If you're standing hunched over the off-axis guider or the guide scope, it becomes nearly impossible.

Don't Drop off your Astrophotos

Your picture of the Great Orion Nebula is very different from the kind of pictures your local photo processor usually gets. But your local minilab proprietor can do a fine job if you tell her or him what to expect. What you should do is pick a slack day or time to bring in your film. An evening when the minilab technicians are standing around bored is perfect. In addition to your film, bring a print of a similar object with you to illustrate what you want. Tell the tech what you've got and show him your print as an indication of what you want. Be friendly and not demanding. Most photo lab personnel are interested in photography and will look upon your pictures as a challenge. During the time of Comet Hale-Bopp, my local lab did a superb job with my astrophotos. The owner was so impressed with my shots he offered to make me a series of 8 × 10s in exchange for being able to display one of the comet photos in the shop window (he also told interested people who they could buy a print from!).

Observing

A Dark Hood

The difficulty in seeing deep sky objects from urban and suburban sites does not come only from the general brightness of the sky, but also from nearby ambient lights that prevent your eyes from becoming even partially dark adapted. You can rig up a series of light shields to protect your telescope from the glare of a neighbor's security light, but this becomes impractical if you have to move your scope around to avoid trees, etc. A good solution is a dark hood. This is a piece of dark cloth large enough to go over your head and eyepiece and block stray light. Picture the dark cloths old-fashioned photographers used on the backs of their cameras to shield their focusing screens from bright light.

A piece of black rip-stop nylon or other similar light but opaque black material will work well. You may think you look a bit ridiculous with a dark cloth draped over your head, but you'll be able to make out much dimmer objects by using this hood. A hood also helps if you're using light pollution reduction filters. Stray light coming in from the eye end of your eyepiece and reflecting off the filter's surface can greatly reduce its effectiveness.

An Eye Patch

An eye patch can serve two purposes. Place one over your dominant telescope eye when you must go inside to retrieve an accessory or answer the phone. This will preserve dark adaptation. Just be careful not to bump into furniture – without the use of both eyes your depth perception will be lost. At the telescope, an eye patch can be used to cover the eye you don't use for observing. Most experts will tell you that keeping both eyes open while observing just works better. Squinting one eye can become tiring and adversely affect your ability to "see." If there's a lot of ambient light around, though, most people find it difficult to keep the other eye open. Put an eye patch over this eye and you can keep it open in comfort while observing. And yes, you will look a bit funny wearing this. Your author with his long hair, beard, and eyepatch can easily be mistaken for an old time pirate, especially when he has that wild look that comes from chasing a dim galaxy cluster!

Binoculars Help

No matter how good your finder technique, using binoculars to examine the area of the sky harboring your deep sky target can help you get the finder in the right place. Looking with binoculars rather than squinting through a finder while hunched over the scope can give you a better idea of the lay of the land. Just remember when mentally transferring the location of your target from binoculars to finder that the binoculars give an upright image while the average finder scope gives an inverted one.

Rod's Rule

Remember *Rod's Rule* when trying to determine the orientation of the field in your main scope, your finder, or another telescope. A telescope with an *even* number of mirrors (this includes telescopes with no mirrors, as a refractor used without a star diagonal) delivers an image that is inverted – upside down – but mirror-correct. A Newtonian reflector is another example of this arrangement, as is your CAT when used without the diagonal. A scope with an *odd* number of mirrors – a refractor using a star diagonal (one mirror) or an SCT with a diagonal (three mirrors) – presents a field that is right side up but *mirror-reversed* from left to right.

Look Away – Averted Vision

Because of the way the human eye's built, looking slightly off to the side of a dim object rather than straight at it will allow you to see it much better, bringing the eye's dim-light receptors fully into play. This use of averted vision can be just as rewarding from light-polluted backyards as from dark sites.

Jiggle-Jiggle

The human eye also finds it easier to detect moving dim objects than stationary ones. This facility once no doubt helped us avoid stalking leopards. These days it can be tremendously useful in tracking down that wild Herschel 400 Galaxy. This is one time when a mount that vibrates a little when you tap it will come in handy. With the object of interest or the location of a suspected object in the field, give your scope tube a gentle tap in order to start it vibrating. This should make the dim sprite show up a bit better. Slewing it back and forth in the field using a slow-motion control will also help.

Planets from a New Perspective

Looking at a scene from a new perspective will often allow you to see details you overlooked before. Artists often employ this trick when working from a photograph. Turning the picture upside down allows them a fresh view of it. You can do this at the telescope, too, and you'll find this especially helpful when straining for details on the planets. How do you turn the image of a planet upside down (or to any other new orientation)? Simply reposition yourself at your star diagonal. Begin, for example, by looking into the diagonal from the right side of your scope. That is, face the side, not the end of the diagonal from the right of your scope's rear end with your body parallel to the scope tube. When you're ready for a different perspective, move around to the left side of the scope and look in. I find this trick really helps with Jupiter.

Keep Those Footsies Warm

Most observers know they have to keep their heads warm in cold weather. This is a prime avenue for the loss of body heat, and once you start getting cold it is sure you won't be able to concentrate on the dim deep sky objects your scope is delivering. Fewer observers, especially those from southern climes, know it is just as vital to keep your feet warm. You could invest in an expensive pair of ski boots or other insulated footwear, but unless you live where it's really cold, this is probably overkill. A piece of carpet to stand on or rest your feet on if you observe while seated can be nearly as effective as expensive boots. An integral part of my observing kit is a nice, thick bathmat. The rubber covered bottom side of this mat keeps out moisture and helps insulate me from the ground. Couple this with the deep pile on the other side, designed to absorb bathroom moisture, and my feet rarely become cold.

Handsies Get Cold Too

Hands are a problem for the astronomer. They get cold, but covering them with warm gloves makes it hard or impossible to delicately adjust focus or push a small drive corrector button. You could remove your gloves temporarily, but this becomes a nuisance, and if you have to leave one of your gloves off for the duration of a long photographic exposure in order to push hand paddle buttons, you might as well not wear them. What you need are the *right* gloves. Soft and supple gloves made from deerskin can keep your hands nice and warm but preserve your manipulative ability.

You can also purchase a variety of clever hand warming devices that can help a lot. Some of these are in the form of small sealed packets that contain chemicals which, when mixed, generate heat. Start the handwarmer according to the instructions, and the chemicals will begin generating heat. Most warmers will continue to generate heat for several hours. At the end of their life, the little (inexpensive) packets are discarded. Warmers that are heated in the microwave and emit heat for several hours are also available. These are made in a variety of shapes and sizes and are often sold for use by spectators at wintertime sporting events, so check sporting goods as well as outdoor equipment vendors for these.

Hyperventilate?

This may be an urban legend among deep sky observers, but I'll offer it up for you to decide. Many of the astronomers chasing the very faintest DSOs claim that taking several very deep breaths will oxygenate your blood and will enable you to see dimmer objects. I've tried this on occasion, and thought I noted a very slight improvement in my view. But this improvement was small enough that it could have been psychological in nature. Do not, of course, take so many deep breaths that you really hyperventilate and pass out! Especially not if you're observing through a friend's giant Dobsonian telescope and are at the top of a tall ladder.

Bilberry?

Bilberry tablets (the bilberry plant is related to the cowberry plant, if that helps), which contain an extract of the plant leaves and fruit, are said to help your night vision. Legend has it that RAF pilots took bilberry for night missions during the Battle of Britain to increase their ability to see faint objects. True or urban legend? I haven't tried this yet, but some in-the-know deep sky observers swear by these pills. If you'd like to try this, bilberry tablets are available in most health food stores. A check of a bottle of the pills found at my local mall showed that "increases night vision" was noted on the label as one of the *supposed* benefits of this substance. Make sure before trying bilberry that its use is not contraindicated by any medical condition you may suffer from. You must also take the pills for a couple of weeks, supposedly, before you'll notice any improvement in your night sight.

Don't Smoke

Ever. Smoking lowers oxygen levels in your blood and really does have a deleterious effect on your ability to see faint objects. I smoked for 20 years before deciding to quit when I reached age 40. I was surprised at just how much my dim object perception improved. Much more than could be attributed to psychological reasons. If you can't quit completely, at least try to avoid smoking a few hours before observing. Your neighbors at a star party will definitely appreciate your abstaining while on the observing field too.

Don't Drink Either (until Later)

Drinking alcohol also seems to have a bad effect on night vision. Being slightly intoxicated will no doubt tend to hinder your ability to track down hard-to-find objects as well. "One too many" can also lead to dropping expensive eyepieces and other equipment. Drinking alcohol, the medical experts tell us, will make your body colder too (though you'll feel warmer at first), putting a quick end to your tour of Orion. I'll occasionally drink a beer when doing non-critical observation of the Moon or a planet from my home, but I usually save the alcohol for the end of the observing run. Nothing like popping a beer or drinking a nice whiskey when the equipment's all packed away at the end of a wonderful evening. Sipping my drink and watching the Sun rise and the stars wink out is a lovely finale to a night of touring the cosmos!

Chapter 10

The Advanced CAT: Imaging and Computers

Even if eyeballing the cosmos, using your CAT for visual observing, is your only interest at first, as time goes on you may become interested in using your telescope to record what you see. You may also become attracted to the idea of hooking your scope to a computer. Computers and imaging are complicated topics, and what's included in this chapter should be considered merely an introduction to the high-tech side of the SCT. Photography and CCD picture taking, especially, are subjects for entire books. But the following should be enough to get you and your CAT on the path to the complex modern side of amateur astronomy.

Deep Sky Photography

As I said early in this book, many amateur astronomers, especially novices, *claim* they want to take photographs through the telescope. But few of these aspiring astrophotographers ever take picture one, much less produce images to rival those by masters of SCT astrophotography like Jason Ware.

Why? The road to beautiful astrophotos is and always has been a long one. The first hurdle for the new astrophotographer is equipment. Most prospective photographers are taken aback when they realize they'll have to invest in a considerable amount of additional equipment before the first exposure can be made. The SCT is very capable of taking astrophotos, but this does not mean it is photo-ready out of the box. If our budding celestial photographer does decide to take the plunge and acquire the additional gear needed for deep sky photography, he or she is then faced with the "three demons" of astrophotography, three difficult problems that must be overcome if pleasing images are to be produced: accurate polar alignment, accurate focusing, and accurate guiding. But if you really want to take pictures, the equipment required for photography can be accumulated for a reasonable cost, and the three demons can be exorcised.

Equipment

Cameras

New astrophotographers will be well advised to begin their astrophotography odyssey with the ubiquitous 35 millimeter single-lens reflex. These cameras are perfect for astrophotography for a number of reasons. Today's models are lightweight, helping prevent balance problems that can hurt your telescope drive's performance. Their through-the-lens focusing method is helpful – really mandatory – for photography, too. 35 millimeter single-lens reflexes are also comparatively inexpensive, especially if purchased used. The SLR has been the single most popular camera for amateur and professional photographers for the last 30 years, so there is no shortage of used ones. And there is certainly no shortage of SLR models and options. Which brings us to the first problem for the new astrophotographer. Which model of SLR is suitable for use at the telescope? Even a quick glance through a photo magazine will reveal a bewildering variety of cameras.

We can narrow down the field of SLRs suitable for astrophotography immediately by specifying that what we want is a mechanical, non-electronic, non-computerized camera. Why reject the most modern types of SLR? Because of one reason: power. When taking deep sky astrophotos the shutter of the camera must be left open for a long time. Maybe for 15 minutes, maybe for 2 hours – or more. Unfortunately, many sophisticated electronic cameras need battery power to hold the camera's shutter open. On a cold night, it is not only possible but *likely* your high-tech single-lens reflex's small batteries will run out of power and the shutter will close of its own accord before you're ready to end an exposure. You'll also find that many of the modern cameras' niceties like autofocusing, programmed exposures, and complex readouts are totally unneeded for astrophotography. In fact, these things just get in the way. The best choice for celestial photography is a simple camera that operates without batteries.

Suggested New Models

Pentax K1000

Sad to say, fully mechanical SLRs are now in a distinct minority in the photography marketplace. But a few are still out there, often being sold as "student cameras." One of the best and least expensive mechanical cameras of all time, one perfect for the astrophotographer, was the legendary Pentax K1000. Rising production costs led Pentax to discontinue this model in 1998, but if you search around you can *still* find these cameras for sale new by photo dealers. The K1000 is just about perfect for use with your SCT. It is reasonably light, fully mechanical, has a bright viewfinder, and is very reliable. The only drawback to this camera is its lack of an easily accessible method of locking up its mirror.

All single-lens reflexes incorporate a movable mirror that diverts the light coming from the lens to the viewfinder for focusing. When the shutter release is pressed the mirror must pop up out of the way to allow the image to reach the film. This action can cause unwanted vibrations that can ruin your astrophoto. A camera featuring mirror lockup, the ability to move the mirror up before taking the exposure, is desirable.

Other than mirror lockup, the K1000 can perform quite well. The focus screen used to view the image is fairly bright, allowing accurate focusing, a prime requirement for a usable astrophoto camera. The K1000 uses the common Pentax K bayonet-type lens mount, so many different lenses, both from Pentax and third party manufacturers, are readily available.

The Cosinas

The name of the Japanese camera manufacturer Cosina is not one that's familiar to most photographers, even experienced ones. Cosina does not sell cameras under its own name very often. What the company does is produce simple mechanical single-lens reflexes for other camera companies. Currently, Ricoh, Olympus and Nikon sell these sturdy all-mechanical cameras under their nameplates. The Cosina SLRs differ slightly in features depending on which brand you buy, but they are all quite similar. Like the K1000, they do not require batteries to take pictures, only to operate their LED-style light meters. The lens mount used on these SLRs depends on which company a particular camera is sold by. The Ricoh Cosina, the KR-5 Super II, uses the venerable Pentax K mount; the Olympus OM-2000 uses an Olympus mount, and the Nikon FM-10 takes standard Nikon lenses.

These cameras are inexpensive but well built. My own Ricoh version has produced many pleasing astrophotos and has endured a number of trips to the dusty west Texas desert without complaint. One advantage the Cosinas have over the K1000 is that they incorporate mirror lockup of a sort. If you fire the shutter by using the camera's self-timer rather than just pushing the shutter release, the mirror pops up before the shutter fires. Set the camera shutter speed dial to "B", engage the self timer, fire the shutter with a cable release, and the mirror gets out of the way in time for vibrations to die out before the shutter opens.

Nikon FM-2n

Like the K1000 and the Cosinas, today's other readily available mechanical camera, the Nikon FM-2n, only needs a battery to operate its exposure meter. But it is *very* different from the other manual SLRs in terms of quality. The Pentax and the Cosinas are Volkswagens. The FM-2n is a BMW. It is also much more expensive than other current mechanical cameras, costing about three times as much as our other candidates. For this premium price you get a camera built like a tank, ready to take on the most demanding professional assignment in the most harsh conditions, but still operate like a fine watch. It is up to you whether you need quality like this, but the FM-2n is a dream to use. It also offers a significant advantage over the cheaper cameras: interchangeable focusing screens. Special *Intenscreen* focusing screens from the Beattie company can be purchased for the FM-2n. Replacing its stock screen with one of these dramatically increases the brightness of dim objects, making it much easier to achieve perfect focus. The FM-2n also excels as a terrestrial camera and can provide all of the picture-taking horsepower you need for a lifetime.

Used Cameras

Buying used is an excellent way to obtain a camera for use in astrophotography. Used mechanical single-lens reflexes are priced very reasonably and you may be able to get a working camera for astrophotography for next to nothing if you're willing to accept a few defects. If there's anything wrong with an old SLR, it often involves the exposure meter. Either the electronics have gone bad or the batteries required for its operation are no longer available (most older SLRs use now-discontinued mercury cells). A light meter is of very little use for the astrophotographer so you can safely buy a heavily discounted camera advertised as "light meter doesn't work." If you intend to use your camera for deep sky work only, you can even accept an SLR with shutter problems. As long as the B function, the speed that holds the shutter open until you let go of the shutter release, works you can use a camera perfectly well for deep sky photography. Naturally, if you'd like to employ your camera for terrestrial as well as celestial snap-shooting, you should choose a camera with no defects.

But *which* used camera? A visit to your local camera store will undoubtedly turn up a case full of older SLRs for sale. Any of the cameras mentioned in the "new" section above, particularly the K1000 or the Nikon FM-2, the FM-2n's predecessor, are good candidates. Older Nikons like the F and F2 also make great astrocameras. Many long discontinued Pentaxes also work well. But there is one no-longer-made camera that has become the darling of today's astrophotographer, the Olympus OM-1.

What makes the OM-1 so great for astrophotography? Well, it certainly wasn't designed as an astrocamera, but its array of features conspires to make it just perfect for this difficult art. It's small, far smaller than most other cameras of the time (the OM-1 was produced from 1973 to1979), it has the all-important mirror lock-up feature, its focusing screens can be changed, and it is completely mechanical in operation. Look through an issue of an astronomy magazine and it's quite likely most of the best amateur 35 millimeter astrophotos were produced by the OM-1.

What *don't* I like about the OM-1? It's old. This camera has not been produced in over 20 years and is no longer supported by Olympus in any way. The OM-1 is also often overly expensive, considering its age. I don't know whether it's because of camera collectors or astrophotographers, but this is one of the most expensive older cameras on the used market – often selling for more than younger Nikon FM-2s. Whether because of age or design deficiencies, the OM-1 can also be temperamental. I've seen the shutter on more than one OM-1 jam at an inopportune moment. Despite these minuses, many amateurs swear by the OM-1 or the slightly newer OM-2n. And the proof is on display for all to see in hundreds of OM-made deep sky images.

Off-Axis Guiders

The clock drives on modern SCTs are good but not perfect. They'll keep an object in view for as along as you want to look if you're properly polar aligned. But small imperfections in your drive gears will cause periodic error, and this small east–west drift will turn your nice round stars to short lines – trailed stars. You must monitor your drive's performance by observing a star while the picture is being exposed. You'll do this with an illuminated cross-hair eyepiece. If the star drifts out of the junction of the cross-hairs, you'll push the east or west button on your telescope's drive controller to put it back. You may also have to occasionally move the star north or south in declination, but this should be rare if you've done a good polar alignment.

If you're shooting piggyback, that is, if you've mounted your camera on top of your telescope with a bracket and are shooting through the SLR lens rather than the telescope, you insert the cross-hair reticle eyepiece into your main scope, and monitor any star in the field as a "guide star." But how can you observe a star at the same time your camera is taking a picture through the main scope? This is done by using one of two instruments: a guide scope or an off-axis guider.

A guide scope would seem to be the easiest solution. You mount a telescope of some kind (usually a small, long focal length refractor) on your main telescope's tube with mounting rings similar to those used by your finder scope. You point the main scope at the target, and attach the camera. Using the adjustable guide scope rings, you put a bright star in the field of the guide scope without moving the big scope off target. You open the shutter, and guide by looking through the guide scope. Easy, but, especially with SCTs, *not very effective*.

There are two reasons a separate guide scope is a poor choice for SCT photographers: *mirror flop* and *differential flexure*. Because of the moving mirror nature of the SCT's focusing system, the mirror can change orientation very slightly as the telescope moves. You may be guiding an exposure perfectly and still come up with trailed stars because of a little mirror flop. As the drive moved the scope along, the mirror moved slightly at some point, causing the star to move slightly in the field. You didn't see anything out of the ordinary through the guide scope, and thought all was well – but your picture was ruined. Differential flexure is a slight bit of flexing between the guide scope and the main scope. If the tube of one of the scopes moves a little with regard to the other, you'll get trailed stars. What you were looking at in the guide scope and what the main scope was seeing were two different things. Differential flexure can, in theory, be eliminated by the use of very sturdy guide scope mountings, but in practice is hard to cure given the limited mounting options of the stubby fork-mounted SCT tube.

The way to eliminate star trailing caused by these problems is to ensure that what you're guiding on and what the scope is photographing are *exactly the same thing*. If the scope's mirror moves slightly, you need to be able to see it happen. There is only one way to achieve this: with an off-axis guider. An off-axis guider or OAG (see Figure 10.1) is really a very simple little device. Start with a prime focus adapter, a tube that allows you to mount your camera without its lens on the rear port of your scope. This is really just a metal tube with threads for your scope rear port at one end and for a camera T-ring adapter at the other. Add to this a small prism and an eyepiece tube mounted

Figure 10.1. An off-axis guider allows you to view a guide star and take a picture at the same time (photo courtesy of Meade Instruments Corporation).

perpendicularly to the prime focus adapter's tube and you have an off-axis guider. What happens is the small prism intercepts a bit of light at the *edge* of the telescope's field of view and diverts it up the eyepiece tube to a reticle eyepiece. You're able to monitor exactly what the main scope is looking at and adjust the drive rate if anything – periodic error or mirror flop – makes the SCT's aim change.

Are there any difficulties associated with using an OAG? Only one major one, the difficulty in finding appropriate guide stars. With a guide scope you can choose just about any star in the general vicinity of the main scope's aim. The ring mount gives you the ability to aim the guide scope fairly far from the main scope's direction in search of a good guide star. But with an off-axis guider you're limited to whatever stars you can find around the periphery of your scope's field of view. Since the edge of the field of an SCT is where images are of the lowest quality, these stars may look more like *blobs* than pinpoints, too. Often you must change the telescope's aim slightly to be able to find any guide stars at all, compromising the framing of your picture.

How do you get a guide star in the field of your OAG? The simplest units allow only one simple motion. You loosen the lock ring that attaches the OAG to your scope, and rotate the guider radially, scanning around the edge of the field for a star. Some of the more expensive OAGs may also allow you to move the prism inward, farther into the light path.

And how about that? Doesn't having a prism in the path of the scope's optics ruin the picture? No, not normally. The prism is usually just outside the 35 millimeter frame and doesn't cause many problems. It is true that if you have the prism in just the right orientation with regard to the film frame you may be able to detect its presence in your pictures. Usually, the shadow of the prism looks like a slight darkening of the image in one of its corners or along a vertical edge. I rarely notice this effect unless there is some sky fog on the negative or the field is filled with nebulosity. Normally any prism shadow blends in with the dark sky background. Both of the major SCT makers and several independent companies are now making off-axis guiders for SCTs.

Guiding Eyepieces

As with any other eyepiece, the most important consideration in selecting a guiding eyepiece is the basic quality of its optical design. Choose a good Orthoscopic or Plössl rather than an inexpensive Kellner. You'll be doing a lot of looking through this eyepiece, and the sharper it can make those already compromised edge-of-field stars the better. You'll also need to select a cross-hair style. An eyepiece with a reticle similar to the cross-hairs in your finder isn't very easy to use. It's hard to tell exactly which way a star is drifting with only two crossed lines. Much better is a double cross-hair reticle where twin lines cross in the center of the field, creating a small box. All you have to do is keep that pesky guide star in the little box (easier said than done). Some photographers like bullseye style reticles even better. With these you keep the star in a small circle in the center of the field. Avoid eyepieces with overly complicated reticle designs – these can be distracting and illuminating them can generate enough light to obscure dim guide stars.

You'll also need to settle on a focal length. Guiding eyepieces are commonly available in focal lengths of 12 millimeters to about 6 millimeters. Normally you should choose the shorter focal length. To ensure accurate guiding, you'll need as much magnification as possible. You may need to add a Barlow, even with a short focal length eyepiece. 300× is not *at all* outrageous as a guiding power for use by an experienced photographer.

How do you light up the cross-hairs? Just like those in your finder, the cross-hairs in a guiding eyepiece will tend to disappear under dark skies. But it is critical that you be able to see them clearly enough to tell when your guide star begins to drift – even slightly. Astrophotographers use small screw-in illuminators for this purpose. These consist of a red LED powered by small button type cells. There will also be a control that will allow adjustment of the brightness of the LED. If your telescope has a power outlet for a reticle eyepiece, you may choose a "wired" illuminator designed to plug into this port. This would be nice since the small batteries that power illuminators don't last very long at all, even just powering a single small LED.

Declination Motors

A telescope that is accurately polar aligned shouldn't be called on to make many north–south corrections during photography. But it is possible you may have to guide occasionally in declination due to minor misalignment, mounting flex, or other problems. It is also often desirable to be able to move the scope at slow speeds in declination to initially place a high-magnification guide star in the cross-hairs of your off-axis guider. To do either of these things, you'll need a motor for your declination axis. Trying to adjust your declination slow-motion control by hand will almost inevitably cause vibrations that will ruin your picture. Many of the most recent fork-mounted SCTs come equipped with built-in declination motors. If your telescope is not furnished with one, adding one to your CAT is simple and relatively inexpensive.

If you're the proud possessor of a GEM mount, you may already have a motor on your declination axis, *if* you had the foresight to buy a dual axis drive when you bought the scope, that is. If you only purchased an RA drive system, I hate to say it, but you may be faced with buying a whole new drive for your scope, including a new hand controller and right ascension motor. Some single-axis GEM drives may be upgraded to dual axis, but most will at least require the addition of a replacement hand controller if not a whole new system.

Wedges

Every fork-mounted SCT comes with a wedge? Right? That used to be true, but no longer. Today's computerized CATs can track the stars without being polar aligned, so both Meade and Celestron have made wedges optional items. You can do almost any kind of visual observing with a computerized scope in alt-azimuth mode, but for any type of imaging you must have a wedge. A telescope that is not aligned on the Celestial Pole will show trailed stars in photos no matter how well it tracked. If you need to buy a wedge for your scope, both Meade and Celestron (as well as a few third parties) sell them. Usually you may choose from either a standard wedge or a heavy-duty or super model. You may be able to get by with a standard wedge for a light 8 inch SCT, but for a larger aperture scope or a heavy 8 inch, a heavy-duty wedge is highly recommended. Much of the vibration present in SCTs, surprisingly, comes not from the fork-mount but from the wedge.

Altitude and Azimuth Adjusters

In order to accurately polar align your telescope for photography, you'll have to be able to move the scope by small amounts in altitude and azimuth. If your wedge does not include fine-adjustment controls, you'll want to purchase one of the add-on kits furnished by the manufacturers. Lighter SCTs may be able to get by with the latitude and azimuth adjusters that came with the wedge, but even a telescope as lightweight as an LX-10 is much, *much* easier to polar align with a fine-adjuster kit installed on the wedge. In my opinion, these items are mandatory for photography.

An f/6.3 Reducer–Corrector

You can take photographs at your SCT's native f/10 focal ratio, but this slow speed makes for punishingly long exposures of all but the brightest deep sky objects. An f/6.3 reducer–corrector makes a huge difference. The nebula that took an hour to record adequately at f/10 is now well exposed after 20 minutes! If you're guiding your scope manually, you'll really appreciate anything that makes exposures shorter. An added benefit of the r/c is it gives you a substantially wider field, expanding the number of objects you can do justice to. An r/c, you'll recall, helps flatten the field of the SCT too.

Film

You certainly can't take pictures without film. But not any old film will do. You need an emulsion that performs well in long-exposure deep sky photography. There are several requirements. First, the film should have good reciprocity characteristics. All films suffer from something called *reciprocity failure* to some extent. Once an exposure extends longer than a few seconds (which it almost always will if you're photographing the deep sky), it takes longer and longer for the density of the image to increase as time goes by. In other words, making a 2 second exposure twice as heavily exposed (dense) may take much longer than an additional 2 seconds. Reciprocity failure in effect lowers the speed (ISO/ASA rating) of the film you're using.

Another very important factor in choosing astro film is its spectral response. Films, both color and black and white, respond differently to different colors of light. There is a tendency for many photographic films to be more sensitive in the blue end of the spectrum than in the red. This is fine if you're interested in photographing star clusters or galaxies where most of the light tends to lie toward the high-energy end of the spectrum. But a film that is unresponsive to red light will spell disaster if you're trying to take a picture of a diffuse nebula, as these objects radiate most strongly in the red light of hydrogen.

What type of 35 millimeter film meets the astrophotographer's requirements as far as speed/reciprocity characteristics and spectral response? If you're interested in making color prints, you're lucky. Many of the color print films on the market today do very well in both respects. Some I have found to be very good are Fuji's 800 speed Superia print film, Fuji's 400 speed NHG professional film, and Kodak's 800 speed Gold film. There are several other excellent color films too – talk with your area's astrophotographers and find out what they're using. But the above are readily available in almost any location. One caution: film companies are in the habit of changing the formulas of their emulsions frequently lately. What was once a good film may be changed suddenly and be terrible for astrophotography in its new form. An example is Fuji's 400 speed amateur color print film. It was formerly the film of choice for color deep sky prints. Unfortunately, Fuji saw fit to "improve" it with the result that it is now insensitive in red and useless for nebulae shooting.

Want color but prefer slides? There are several reasons to do transparencies rather than prints, but the major advantage is that the processing is standard – you won't have to hold the photofinisher's hand when you have your pictures processed. It is a simple matter to have prints made from your good slides by a lab, too, so don't worry about being forced to display your work with a slide projector. You many even be able to do this at home, as many modern scanners will allow you to make prints from slides by scanning your transparencies into the computer.

Black and white *is* beautiful. There is a depth and richness to black and white images no color print can match. But black and white is something of a problem for the deep sky photographer. The first stumbling block is you'll probably have to have your own darkroom setup to process black and white film and make prints. If you live in a major city, you may find a lab that does black and white, but in most areas this is not an option. It is easy to print and process your own black and white film, but considerable time and expense is involved in putting a darkroom together and learning how to use it.

Then there's the film issue. There is really only one good black and white astrophoto film: Technical Pan from Kodak (also known as TP2415). But you can't just run down to the camera store, buy a roll and start taking astrophotos. As sold, Tech Pan is very slow and unsuitable for deep sky shooting. It must be *hypered* before shooting. Hypering is the process of baking film in a bath of hydrogen gas for hours to eliminate reciprocity problems. Hypered Tech Pan is incredible for astrophotography; it produces deep, rich images where the red light of nebulae is readily recorded. How do you get your Tech Pan hypered? You can buy it pre-hypered from astrophotography merchants like Lumicon. But buying rolls of hypered film is expensive and inconvenient. Your own hypering system is another solution – you can build or buy a hypering chamber. This is something best left for advanced workers, though. A heated chamber filled with hydrogen "forming" gas is nothing for novices to monkey around with in the kitchen.

What about the other black and white films? While many venerable emulsions like Kodak's Tri-X Pan and Plus-X Pan can be used for lunar and planetary photography, they lack red response and suffer from relatively poor reciprocity characteristics, making them undesirable for photographing the deep sky.

T-Rings

How do you attach your camera to the off-axis guider? You'll use a T-ring. A T-ring is a machined metal ring with T-style screw threads (similar to the old Pentax screw-thread lens mount) on one end and a shape that is designed to fit your camera's lens mount on the other. You remove your camera's lens and attach the T-ring in its place. You can then thread the camera/ T-ring combo onto your OAG. All off-axis guiders and other camera adapters feature standard T threads. This means that once you've equipped your SLR with a T-ring you can put it on any photo accessory sold for your scope.

Cable Release

A cable release (or remote shutter release) allows you to gently fire your camera's shutter and to hold it open in the B mode. A cable release is equipped with a lock that keeps the shutter release depressed when desired. You can buy cable releases with set-screw type locks, or automatic locks that engage every time you press the cable release. To unlock, you depress a lever or ring on the end of the cable. I prefer this auto lock type.

The Astrophotographer's Three Demons

Astrophotography just ain't easy. There are so many things that can go wrong out in a field in the middle of the night. Most astrophotographers are happy to get *one* good image per roll of film. There are three areas that cause the most failures, though, and by concentrating on banishing these three "demons," you can maximize your chances of at least getting that single beautiful print per roll. These three fearsome creatures are the Demon of Polar Alignment, the Demon of Focus and the Demon of Guiding.

The Demon of Polar Alignment

Those much labored-over deep sky shots pour out of the envelope at the photolab. At first you're ecstatic! They're beautiful! But closer examination reveals a problem. The stars in some of the shots are not *exactly* round points of light. Stars near the edge of the frame, especially, look like little curved arcs. The stars actually seem to be rotating around a central point in the frame. This point is your *guide star*, the only really perfect star image in your photo. You guided wonderfully well, but poor polar alignment caused field rotation. The only cure for this is to improve your polar alignment technique. You must get the RA axis of your telescope pointed straight at the North or South Celestial Pole. Just aligning on Polaris isn't good enough. Even the fairly accurate alignment produced by polar alignment finders is usually not precise enough for exposures more than 5 or 10 minutes long. What is needed is a *drift alignment.*

Drift alignment is the process of accurately adjusting a telescope's mounting by watching the north–south and east–west drift of selected stars. This procedure will seem a little complex at first, but after doing it a few times it will become second nature. Sure, it takes time, but it is much better to spend half an hour adjusting your CAT's wedge or GEM mount than hurrying through polar alignment, spending an entire night taking pictures, and finding that they are all ruined by field rotation when you have them developed. These instructions are tailored to the needs of fork-mount users, but should also be usable on those scopes equipped with German mounts as well.

Begin your drift alignment by doing a good normal alignment. Point your RA axis at the pole as accurately as you can using your method of choice. This can be a polar alignment scope or merely the simple "point the

RA axis at Polaris" method we've used earlier. The closer you get to the pole initially, the quicker and easier your drift alignment will be.

Your initial task when doing a drift alignment is to locate a star that is close to the meridian (the imaginary line that runs through the North and South Celestial Poles and the zenith) and around 15–20 degrees north of the Celestial Equator (i.e. at a declination of about +20 or so). This is not hypercritical. Just find a nice medium-bright star near this general location. If you're not sure you're in the right area, reference your star atlas. Find the line that indicates the Celestial Equator. Choose a constellation close to it that is almost overhead and pick a star from this constellation about 20 degrees north of the equator line.

When you have the star in the cross-hairs of an illuminated reticle eyepiece at about 250× (use a Barlow if necessary to achieve this magnification) rotate the eyepiece barrel in the diagonal until the cross-hairs are aligned N/S and E/W. That is, until the star moves nicely along the vertical cross-hair when you turn the declination slow-motion knob and along the horizontal one when you turn the RA slow-motion control.

Now let's *drift*. Put the star in the center of the cross-hairs using either the electric or manual slow-motion controls and watch for movement *up or down* – for drift along the N/S cross-hair. Don't worry about the E/W drift of the star in RA. You can guide in RA with your hand controller to keep the star near the center of the cross-hairs if you wish, but all you want to know at this time is whether the star is drifting *up or down in the field*. If your alignment didn't put you dead-on the celestial pole (it probably didn't, but it *does* happen once in a while), you'll detect some declination drift after about five minutes or less. If you see that the star drifts *up* in the field, adjust the azimuth (left–right motion) of your wedge to make the star move *right* in the field. How much? About the same distance it has drifted. If the star drifts *down*, move the star *left* in the field with your azimuth adjuster. Once you've adjusted the wedge in azimuth, use the RA and declination controls to recenter the star in the eyepiece. Keep watching for up/down drift and moving the wedge in azimuth until you can go *at least* 5 minutes with no drift. *At all.*

An accurate drift alignment also requires you to adjust the altitude of your wedge. We'll use a second star to do this. Locate another medium-bright sparkler.

This time right *on* the Celestial Equator and only about 15–20 degrees above the eastern horizon. Set your eyepiece up as above, with the up–down cross-hair defining declination movement and the left–right cross-hair defining RA. Center the subject star in the cross-hairs and watch for drift. *Once again, we're watching for up–down drift.* But the wedge *adjustment* differs. If the star drifts up, adjust the wedge elevation control to move it *down* in the field. If the star drifts *down*, move it *up*! Keep doing this (recentering the star between adjustments) until there's no visible drift *for at least five minutes.*

Once you've eliminated the drift, you're finished. *Carefully,* very carefully, tighten down the wedge's bolts to lock it in altitude and azimuth. If you've followed these instructions carefully, you should be able to expose for at least an hour with no field rotation. The only variables will be the quality of your guiding and focusing.

The Demon of Focus

Many visual observers find it difficult to achieve sharp focus on dim deep sky objects when using a high-quality eyepiece. Imagine how much more difficult it is to get a faint nebula, galaxy, or star cluster sharp when you're required to look at it projected on the dim ground glass screen of a 35 millimeter single-lens reflex camera! If you're very lucky, there may be a bright star or two in the same field as the object of interest, making focusing a little easier. But it is *still* tough. You can try one of the bright replacement focusing screens for your camera if it's able to accept these, but you may find they're bright *in name only.* And it is *very* critical that your scope be exactly in focus when taking pictures. In visual work, your eye can easily adjust to make a slightly out of focus image look nice and sharp. But even minor focus errors in photography will make your shots look terrible.

There are a number of ways of obtaining sharp focus for deep sky astrophotography. Some involve devices that take the place of your camera at the focal plane. Others rely on special off-axis guiders that include focusing eyepieces as well as guiding eyepieces. Many of these devices work well, but all have two things in common: they cost additional money and are difficult for beginners to use. The method I use for focusing costs nothing and is easy for the freshest novice to get the hang of.

To use my method of focusing you must first accurately focus your camera in exactly the setup you intend to use for your picture taking. If you're going to be using an f/6.3 reducer–corrector, for example, you should go ahead and attach it to the telescope before beginning. With your equipment setup ready, point your telescope and attached camera at a bright star and bring it into sharp focus in the viewfinder of your SLR. Take your time and do this right, as the rest of the evening's focus accuracy will depend on what you do at this point. Strive to make the star as small as possible.

The camera is now accurately focused and you could just point the scope at your deep sky subject and start your exposure. But this is where the trouble begins. It is just about *impossible* to find objects by looking through the camera's dim viewfinder. If you have a goto scope or accurate digital setting circles, this may not be a problem. But if you've got a manual telescope you need a way to find and frame your galaxy or nebula easily. To do this, remove your camera and off-axis guider from the back of the scope. *Be careful not to accidentally turn the focus control.* Screw your visual back and diagonal onto the rear port. Make sure that the diagonal is snugged up into the visual back as far as it will go and that the setscrew is nice and tight.

Insert an eyepiece of your choice, nominally a low-power one. Take a look at the bright star you used to focus the camera. Is it in focus? Probably not. But *do not* touch the telescope's focus control. Instead, bring the star into focus by sliding the *eyepiece in and out of the diagonal.* If your eyepiece will not come into sharp focus no matter how you slide it in or out, try another. Chances are one from your eyepiece collection will come to focus. If necessary, loosen the set-screw that holds the diagonal in your visual back and slide the diagonal itself out a bit. Once you have good focus, tighten down the visual back's and the diagonal's set-screws. Just remember not to move the telescope focus control.

You can now locate your target as you normally would, viewing through the eyepiece. If you've proceeded carefully, you can then remove the diagonal and visual back and replace your camera and off-axis guide on the back of the scope. The camera should still be in perfect focus. After taking your picture, you can remove the camera from the scope, put the diagonal back on, and point the scope at your next subject with ease. It might not be a bad idea to occasionally point the scope at a bright star with the camera in place to check the focus just to be sure all is well.

The Demon of Guiding

Your mounting is exactly polar aligned and your OTA is sharply focused. But there is still one more hurdle: properly guiding the telescope for the duration of the exposure. For some people this is the hardest part of astrophotography. Lately, quite a few photographers have begun using a device that automates guiding. An autoguider is essentially a CCD camera with special software that looks at a guide star through either an off-axis guider or a guide scope and issues guiding commands to the telescope drive when it senses that the star has wandered from its appointed position. Autoguiders make very long exposures practical, but they are still fairly expensive and provide one more layer of complexity for the beginner. Guide manually in the beginning.

Before a deep sky exposure can begin, a guide star must be located and placed at the center of the cross-hair eyepiece in the off-axis guider. This is where many budding astrophotographers run into their first hangup – they just can't find a guide star. There is no denying that this can be difficult. The fact is that an off-axis guider will always present comparatively dim and misshapen star images in the eyepiece. It has to. If the prism were big enough or extended deeply enough into the field to deliver a big selection of sharp and bright stars, it would interfere with the light going to the camera.

Don't give up, though. Keep looking for stars and you'll probably find one good enough to guide on in practically *any* field. It may be unavoidable that you have to move the scope a little to find a good guide star. This will mess up the framing of your subject, but it is better than not getting the shot at all. As you gain experience in astrophotography, you'll be surprised to find you can reliably guide your scope with very dim stars.

Once you've got a guide star in the cross-hairs, get comfortable. Taking a deep sky photo will require that you keep your eye glued to the guiding eyepiece for the duration of the exposure. If you're uncomfortable you'll never be able to do this. Find a decent observing chair of some kind, position yourself in as relaxed an attitude as possible and begin your exposure. In the beginning, it's wise to start with short-duration runs. At first, a short 15 minute run will seem like an eternity!

You're ready to go, but exactly how do you *do* it? Guide, that is? You watch your star very closely. When it moves out of the little box, or bullseye, or whatever other device you have in the center of your guiding eyepiece, you press one of the buttons on your telescope hand controller to move it back. Make this easier by orienting the control in your hands so that the buttons are oriented the same way as the directions of your field of view. You'll find it advantageous to use as high a magnification for guiding as possible, even if this means using a Barlow lens with your reticle eyepiece. The idea is to keep the star in the center at *all* times if possible, and high power will make this easier – you'll see the slightest movement of the star right away.

Guide as well as you can. At first you'll make mistakes. Your attention will wander and that silly little star will leave the cross-hairs before you realize it. But don't quit. Keep the exposure going, even if you're pretty sure you've misguided badly. Like many other things, guiding well is a product of practice. As you gain more experience, it will become easier; you'll get to know your telescope's drive to the point that you'll have a pretty good idea when periodic error will start moving the star off target. At this point you'll find you'll be able to start extending your guiding time long enough to get good exposures of even faint objects.

Site Selection

You've got all the equipment you need and at have least some idea of how to use it, but you also need an adequate site to take pictures from. By "adequate" I mean *dark enough*. The enemy of the deep sky astrophotographer is *sky fog*. A sky that is bright with light pollution will result in an image with a bright, light-fogged, background that blots out your beautiful deep sky object. If your sky's limiting magnitude is about 4.5, typical for many suburban areas, you'll find that sky fog will ruin your images before you've made an exposure long enough to record much. Try to photograph from areas that are as dark as possible – and also secure. When you're guiding photographs you need to relax and concentrate. If you don't feel safe at a site you'll look up from your guiding eyepiece every time you hear a twig snap, and eventually you'll misguide and ruin your picture.

What if you can't *find* a good dark site? Or one that's secure? You can still do some nice work, on brighter objects anyway. You can partially alleviate sky fog problems by foregoing the use of your f/6.3 reducer–corrector and photographing at the normal f/10 speed of your SCT. You can expose long enough at f/10 to record M42, for example, even under fairly light-polluted skies, before the sky fog affects your image badly. The trade-offs are that you won't want to try really dim objects, as the exposures would be *exceedingly* long, and your field of view on the film frame will be much narrower. Shooting at f/10 from your backyard is a good way to keep in practice between visits to a dark site, though. It also allows you to get pictures when there's a Moon in the sky. I've been able to get some pleasing portraits of the Orion Nebula on Full Moon nights by shooting at f10.

A Typical Deep Sky Photo Run

The purpose of this section is to give you an idea of how the usual astrophotography session goes.

Arriving at your site, your first task after unloading the SCT and setting it up as normal is to achieve a good polar alignment. As soon as you can see Polaris, you get it in the cross-hairs of your finder scope or in its proper position on the reticle if you're equipped with a polar alignment finder. After you've used your finder or polar finder to roughly align the scope, you begin your drift alignment, adjusting your wedge or GEM mount until there is no more detectable up/down drift of the star. Drift alignment has taken about half an hour, and it's now dark enough for you to locate your first target.

Referring to the list you've made beforehand, you see that the first object for the evening will be M42. You've photographed this spectacular nebula many times before, but you're still questing after the perfect shot. It is also easy to photograph and will provide a good "break-in" target for the evening. You find M42 and

Figure 10.2. A CAT in the hands of a master can produce amazing photos: Galaxy M51 by Jason Ware (image by Jason Ware with 16-inch LX-200, Courtesy of Meade Instruments Corporation).

place it in the field of the scope. It's looking good tonight. Nebulosity overflows the field! But you don't look at it for long; it's time to start imaging. You attach your loaded camera to your off-axis guider via a T-ring, screw the whole thing onto the rear cell and insert your illuminated reticle eyepiece into the guider's focuser. There are several bright stars visible in the camera's viewfinder when you take a look, so you are able to focus without much trouble. If you were not able to focus easily, you would have moved the scope to a nearby bright star for focusing purposes.

Once you're satisfied that the image in the viewfinder is as sharp as you can make it, it's time to find a guide star. This is a task you normally dread, but you recall that there are a number of good candidates around the edges of the M42 field, so you're hopeful this won't take long. Loosening the lock ring that holds the OAG and camera onto the rear port, you look through the guiding eyepiece to see if there's a candidate star visible. If not, you slowly rotate the guider body around the back of the scope while peering through the reticle eyepiece. Eventually a bright star pops into view. It looks perfect and you only have to move the telescope's mounting slightly to get the guide star into the cross-hairs. A glance through the camera viewfinder shows that the nebula is still nicely framed.

You now ready your camera for the first exposure, attaching a cable release, setting the shutter speed dial to B, cocking the shutter, and setting the self-timer or mirror lock-up lever. Before opening the camera shutter, you arrange your chair, properly orient your hand controller and try to get as comfy as possible. It is wise to guide on your star for a few minutes before opening the shutter so that you can get into the rhythm. When you feel ready, you push the cable release button and lock it. With the shutter open, you spend the next half hour single-mindedly doing your best to keep that star in the middle of your cross-hairs.

When the kitchen timer you use to time exposures dings, you unlock the cable release allowing the shutter to close. Whew! You relax for a moment, but not too long. Between the setup time, drift alignment time, and the exposure itself, you've been on site for a couple of hours now and have only one shot "in the can." There's no time to waste. You either prepare to point the scope at a new target or take a second "insurance shot" of M42. If you move on to a new object, you carefully remove camera and guider from the back of the scope and replace them with a diagonal and eyepiece without touching the focus control. This will save you from having to refocus on your next object. And you just keep going, for as long as darkness and your stamina hold out.

Exposure

How long should you expose each astrophoto? Go as long as possible at first. In the beginning, this will likely be no more than about 20–30 minutes, which is long enough to do justice to the bright Messier objects which should be your first subjects. When you become skilled and hardened enough to guide for an hour or more you can worry about matching exposure time to object (and sky) brightness!

Other Types of Astrophotography

The astrophotography method we've been outlining, prime focus through-the-telescope deep sky imaging, is the holy grail for amateur astronomers interested in picture taking. But it isn't the only thing you can do with your telescope and a camera. And it isn't even necessarily the best starting point on your astrophotography odyssey. Many experts will tell you that piggybacking is the real place to begin. You may also want to try imaging those *other* sky objects – the Moon and planets.

Piggyback Photography

Is exactly what it sounds like: your camera rides piggyback on top of the telescope, making use of its drive to track the stars but shooting though its own lens, not through the main telescope (see Figure 10.3). Piggyback photography is often recommended for beginners since it eliminates or at least reduces the need for precise guiding. Inaccurate polar alignment is

Figure 10.3. For piggyback exposures, your camera rides on top of your telescope on a special bracket and shoots through its own lens.

also much less noticeable when doing piggyback photography than it is through your scope at high magnification.

The only thing you need to get started in piggyback photography is a bracket that allows you to mount your camera on the tube of your scope. These are available from astronomy accessory dealers, are inexpensive in their most basic forms, and mount easily in the accessory holes on the rear cell of your CAT. You might also want to pick up one additional item, a ball-type camera mount. If you place your camera directly on the piggyback bracket, you'll see you're pretty limited in the directions it can be aimed independently of the main scope's direction. Attaching a ball and socket mount to the piggyback bracket and the camera to this unit allow you complete freedom to point the camera around the sky without regard to where the scope itself is aimed.

For your first piggyback shots, start with a normal lens on your camera, a lens with a focal length of around 50 millimeters. This will not only give you a wide field of view, it will also completely eliminate the need to guide the scope if you've done a fairly good polar alignment. With the scope setup and tracking, mount the camera on the piggyback bracket, and point it in the general direction of a photogenic part of the sky using the ball head. Focus the camera as accurately as you can (don't just depend on the infinity mark on your lens) and open the shutter by setting the speed to B and locking it in place with your cable release.

Your normal lens probably has a speed of about f/2 or faster, so sky fog will set in in a big hurry unless the skies are dark. Keep your exposures to a few minutes if the sky is not pristine. Most SLR lenses will deliver maximum sharpness if you stop them down an f-stop or two, so you may want to try a couple of stopped down shots on your initial roll of film and compare the results with wide open frames. If you've got longer lenses, you may try them as well. If you use a lens with a focal length of 200 millimeters or more, you'd be advised to guide the scope for best results, though. Do this by inserting an illuminated reticle eyepiece into the diagonal and keeping a star in the cross-hairs just as in prime focus work. It's really easy to find a guide star when you're looking thorough the main scope rather than an off-axis guider!

The Moon

The Moon is a natural for the beginning astrophotographer. It's beautiful, detailed and bright enough so that almost anybody can bring back a pleasing picture. You don't need any special equipment to get started, either – not even a camera adapter. You just use the *afocal method* of photography at first. To shoot afocally, the lens stays on your camera and the eyepiece stays in the telescope. All you do is point the camera at the eyepiece – it takes the place of your eye – look through the viewfinder, focus using SCT's or MCT's focuser until the lunar landscape is sharp, and snap away!

You can hand-hold your camera or set it up on a tripod next to the telescope. Exposure? Try a variety of exposures, but with this bright object, it's likely that your SLR's exposure meter will work, providing at least a starting point. When your pictures come back, you'll be amazed at how sharp the Moon is! The only problems with the afocal method are that you'll often – depending on camera lens and eyepiece – get a "porthole" effect on your shot. The Moon will be in a circular spot on the film. You'll also notice that as you increase magnification (by using higher-powered eyepieces) it will become impossible to hand-hold the camera and get a sharp picture.

There are two other methods you can use once you grow tired of afocal exposures. The first is prime focus. For prime focus shots you attach the camera to the scope with a prime focus adapter, which is like an off-axis guider without an eyepiece holder or pick-off prism. It allows you to mount your camera, without its lens, onto your scope. The fact that neither the lens of your camera or the eyepiece of the telescope is present means your magnification will be low. But this is nice for full disk shots of the Moon, and is particularly useful for photographing lunar eclipses. If you already own an off-axis guider, there's no need to invest in a prime focus adapter. Just use your OAG. The prism won't show up.

If you want high-magnification images of the Moon, you'll need to use a different sort of camera adapter, an *eyepiece projection adapter* (a *tele-extender*). This is a special adapter tube designed to allow you to project an image onto your camera's film plane with an eyepiece. It works like this: before attaching the eyepiece projection adapter, you put a 1.25 inch eyepiece into the visual back of your telescope. Don't use a diagonal, just put an eyepiece on the scope in "straight-through" fashion. One end of your tele-extender's tube will be specially threaded so that it can be screwed onto the visual back and *over* your eyepiece. The other end will possess threads for a T-adapter of your choice.

With a setup like this, you'll be able to take very high magnification shots of the Moon. The eyepiece will project an image much like a movie projector, producing high magnification. There is a problem inherent in using a setup like this, though: vibration. The magnification is so high your shutter slamming open – even with the mirror locked up – will cause the telescope to vibrate enough to blur the picture. Many experienced lunar photographers get around this by using the *hat trick*.

To use the hat trick shutter method, you'll need a piece of black cardboard large enough to cover your telescope's aperture. With the scope properly aimed and focused and the camera's shutter cocked, cover the corrector plate with your cardboard sheet. Use your cable release to open the camera's shutter (it should be set on B). Now, wait a few seconds for vibrations to die out, and remove the cardboard from in front of the scope for the proper exposure time. Replace the cardboard over the scope and close the shutter. Because the high magnification you'll achieve with eyepiece projection will require long exposures – from around 1 second to several seconds – it's easy enough to act as your camera's shutter. For proper exposure, "bracket", that is, try several different exposure times. Books on astrophotography often include exposure tables for the Moon, and a good book devoted to astrophotography is one of the beginning astro-imager's most important tools.

The Sun

You take pictures of the Sun the same way you do the Moon using the same methods. There is one important difference, naturally, *a filter*. Whether you're observing the Sun visually or photographing it, you'll have to have a filter over the corrector to protect the telescope, the camera and your eyes from severe damage. Some solar filters are sold for photographic use only. They allow more light to be transmitted than visual filters do, so shorter exposures are needed. I advise against using this type of filter. It is all too easy to forget which filter is in place and accidentally look through the scope, resulting in possible harm to your eyes.

The Planets

The planets are the most difficult subject for an astrophotographer. Next to planetary photography, deep sky work is really child's play. The problem for the planetary photographer is that his targets are small. High-magnification eyepiece projection photography is *required* to make the planets large enough to show details on film. The planets become relatively dim at magnifications high enough to make them pleasingly large, and relatively long exposures are needed, even for Jupiter. If you must take long exposures, you must shoot only when the atmosphere is very steady. The slightest amount of "boiling" will obscure planetary features.

Electronic Imaging

You don't need film to take pictures of the Moon, planets or deep sky objects. Increasingly, amateur astronomers are following the lead of professionals and abandoning their film cameras for electronic ones. These cameras, which use electronic chips called Charge Coupled Devices or CCDs rather than film (see Figure 10.4), have long since replaced old fashioned film plates in professional work and are now threatening to do the same thing in amateur imaging. Why? Because CCD cameras have an important advantage over film: sensitivity. The CCD chips in use in amateur cameras can be as much as 100 times more sensitive than photographic film.

The worth of this high sensitivity should be obvious. An image of a galaxy that once required a one-hour exposure with your SLR and your C8 can now be done in a minute or two. No more marathon guiding sessions. Your CCD chip sucks up all the photons you need in this brief exposure and you move on to your next target. It is common for an astrophotographer to be able to obtain more images in a single evening with a CCD camera than a film photographer can bring home

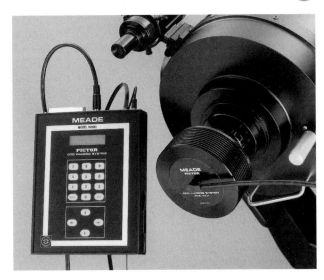

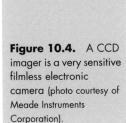

Figure 10.4. A CCD imager is a very sensitive filmless electronic camera (photo courtesy of Meade Instruments Corporation).

in *six months* of photo runs. A CCD camera can also go very "deep," imaging *fifteenth magnitude stars* in a minute or less with an 8 inch telescope. Even distant Pluto falls to the CCD camera in less than a minute from suburban skies. This sensitivity, when coupled with these sensors' excellent dynamic range, makes them much more usable than conventional film in heavily light-polluted environments as well. No more grueling trips to distant dark sites! You'll be able to do at least the majority of your picture taking from your comfortable backyard.

How can an electronic computer chip stand in for film? A CCD chip or *detector* is a special type of electronic component that is fabricated with a light-sensitive surface. This surface is divided into discrete pixels, picture elements, in much the same way as the monitor screen of your computer. The size of a CCD chip and its resolution are expressed in terms of horizontal and vertical pixels. A CCD chip might be advertised, for example, as being 336 × 242. This means that the chip is 336 light-sensitive pixels wide and 242 pixels high. A CCD with a size of 768 × 512 would be larger and would offer higher resolution.

When your telescope is pointed at the sky, photons strike this array of pixels and cause a flow of electrons. The more intense the light, the more photons strike certain pixels or groups of pixels and the more electrons flow. The chip's bounty of photons is counted by the computer when the exposure is completed, and the pixels where more electrons flowed will register as *brighter* than pixels where fewer or no photons were counted. In this way a picture is built up. The camera electronics keep track of the brightness of the pixels and assemble an image based on these results. Most CCD cameras require the services of a computer to make an exposure. A program running on the PC (which may either be an IBM compatible or a Macintosh) controls the camera and saves the images for later processing.

Once the image has been assembled by your camera it is sent to a personal computer for display and processing. At this point, the photographer – usually by the light of day in comfortable surroundings – can play with the image in order to fine-tune brightness, contrast and other factors. This may be done with a program included with the camera or with any type of image manipulation program capable of handling the file format generated by a particular CCD camera. Once the image looks good, it may be printed on one of today's high-quality color printers, published on an Internet World Wide Web page or even made into a 35 millimeter slide or photographic print by a computer service bureau.

There is little doubt that CCDs are the wave of the future for astrophotography. So is it time to junk the 35 millimeter single-lens reflex? Don't be too hasty. CCDs have yet to displace 35 millimeter film in either quality of results or ease of use. Unlike professional astronomers who are mainly interested in scientific data gathering, most amateurs are after *pretty pictures*. CCD camera are now reaching the point where they can deliver attractive images, but to say that the average CCD image looks as good as one developed by the old fashioned film photo process is still laughable. It all comes down to pixels. A huge CCD (for amateurs) might be composed of an array of 800 × 600 pixels. A 35 millimeter film frame, in contrast, is made up of tens of thousands of grains of silver (film's "pixels") in both the horizontal and vertical directions. More pixels make for more resolution and that means finer details and a better looking image.

If you're interested in full color images, the 35 millimeter SLR also wins hands down. The CCDs used in most astrocameras use black and white chips. Color chips *are* available, but they are less sensitive than the black and white models. A CCD user *can* generate color images with a black and white CCD camera by using the *tricolor process*. Three exposures are taken through color filters, one in each of the primary colors, red, blue and green. The computer can be used to assemble these three exposures into one true color image. Taking these three images, properly registering them into one picture, and getting the color balance right, is, as you might guess, work for a very experienced CCD user. Most beginners will be limited to black and white picture taking.

Many novices become enthusiastic about CCDs because they think they will make the astro picture-taking process easier. Although it's true that the time required for exposures is much shorter than for even the fastest films, the CCD picture-taking process is not quite as simple as it might appear. One thing that slows down the CCDer is heat. CCD cameras suffer from a problem called *dark current*. Heat causes the pixels in your CCD chip to falsely generate electrons after only a few seconds of exposure. This appears as noise on your picture – white speckles similar to the "snow" seen on a poor television picture. The only way to reduce this noise is to cool the camera. Most cameras incorporate thermoelectric Peltier coolers that can chill the CCD chip to very low temperatures without the use of liquid or gaseous refrigerants. So the first thing the CCDer has to do after setting up the SCT and turning on the CCD camera is to wait for it to cool off and stabilize. This process can take quite a while depending on the ambient temperature in your location.

Even with the camera chilled, noise will still be present. Some pixels, for example, will be hot: they will always be turned on whether light strikes them or not. This is an inevitable artifact in amateur grade CCD chips. CCD users eliminate this residual noise by taking a *dark frame*. For each image, a frame of identical duration is taken with the lens cap on the telescope. If you're taking a 5 minute exposure of M42 you must also take a 5 minute dark frame with the corrector plate covered. The computer can compare the dark frame and the actual image and remove any hot pixels. Some CCDers only take a few dark frames over the course of an evening, one 1 minute frame, one 5 minute frame, one 10 minute frame and so on. But most authorities recommend a dark frame for each exposure. Some cameras also require a flat-field frame. This is taken with the camera pointed at an evenly illuminated white surface. The resulting exposure is used during image processing to make sure the CCD images are evenly exposed across the frame.

A single short CCD exposure is actually composed of three frames: dark frame, flat field and image. Things aren't quite as quick and simple as it seemed at first. It is true the time involved in three CCD pictures is usually less than that required for one film shot, but the process of arranging the dark frame and flat field shots is actually more involved than taking a simple film photo.

Nevertheless, a 1 minute *unguided* exposure does sound wonderful if your eyes are still burning from all night photo-guiding sessions. And it's true that it's possible to make unguided short exposures if your CAT has a fairly decent drive, especially one that's been trained using periodic error correction. But despite the incredible sensitivity of the CCD chips, many faint objects will still need longer exposures. It will be necessary to guide these. You may find to your dismay CCD cameras actually require *more* accurate guiding than film cameras. This is because of the small size of CCD chips. Many of these detectors are no larger than the nail on your little finger. Even the largest and most expensive CCD detectors used in amateur camera are far smaller than a 35 millimeter film frame. At the telescope this translates into *high magnification*. When processed, an image of a small galaxy will fill the entire frame of a CCD shot. It is much larger relative to the frame than a 35 millimeter image of the same object. This large *image scale* means that your guiding must be much more precise than what you're used to when working at f/6.3 with 35 millimeter film.

The small size/high magnification aspect of most amateur CCD cameras also means that you can forget using your SCT at its native focal length of f/10. At this focal ratio you'll be confined to imaging only the smallest objects, tiny galaxies for instance. This focal length problem can be alleviated by using an f/6.3 reducer–corrector, one of the now available CCD-only f/3.3 reducers, or by shooting with a recent Celestron SCT with the camera mounted in the Faststar position. In this mode the camera is mounted in place of the telescope's special removable secondary mirror and shoots at the SCT's mirror's real focal length of f/2. All of these solutions do mean extra expense for your initial set-up, of course.

Finally, there's the computer question. Some of the most recent CCD cameras can be operated in the stand-alone mode. They can store your pictures in memory for later download to a computer. But most CCD cameras require the use of a personal computer not only for image storage, but even to operate the camera.

Not everyone is willing or able to bring an expensive laptop computer into the field, and an AC-powered desktop machine is not recommended due to safety concerns. A less expensive obsolete laptop computer can be located on the used market, but one that has a fast enough processor to be practical will still constitute a considerable expense.

Have I frightened you away from buying a CCD camera? No? Good! I simply want to emphasize that a CCD camera is not a simple and easy replacement for film-based astrophotography. Should you consider a CCD setup? *Perhaps.* If you're comfortable with computers, enjoy challenges, and are experienced enough with your telescope to be at ease operating it in demanding circumstances, I say "go for it." One suggestion I have for the prospective CCD user is to try prime focus film astrophotography *first,* before spending thousands on computers, CCD cameras and software. Banishing the film astrophotography demons will stand you in good stead in dealing with the similar devilish creatures that lie in wait for the CCDer. I would also hazard a guess that if you find film astrophotography boring, tiresome and overly demanding you won't like computerized CCD imaging either.

CCD imaging really is a demanding and complex field of its own. I have no doubt at all that CCDs will live up to their current bright promise in the next few years, especially if CCD detectors continue to get larger and cameras cheaper. In one area at least, CCD cameras have completely replaced film. Planetary imaging is a natural for these devices. Their high sensitivity means very short exposures can be used, even with the large image scales necessary for revealing detail on the planets. Short exposures minimize the blurring effects of atmospheric seeing. Today's amateur CCD planetary images are of far higher quality than any Earth-based film photography ever has been. Even professional film pictures of the planets taken with huge telescopes are easily bested by amateur CCD images taken with 8 and 10 inch SCTs!

Other Electronic Cameras

Want to improve your lunar and planetary pictures, but don't want to spend a lot of money on CCD equipment right away? You can get your feet wet in electronic picture-taking and computerized image-processing very cheaply. Integrating CCD cameras aren't the only imaging devices that use CCDs and that take advantage of these electronic chips' amazing characteristics. Both the now popular electronic digital still cameras and video cameras use CCD detectors. These devices can be readily pressed into service to take lunar and planetary photos.

Electronic Still Cameras

These are the current darlings of snap-shooters. Imagine a camera that never needs film and can be connected to your computer to allow you to email family snapshots to all your friends. The resolution on these little cameras does not approach that of even a disposable 35 millimeter camera, but their color images *do* look very good. The images these cameras produce are certainly more than good enough to make the average person happy. Many amateur astronomers, upon realizing that these cameras contain CCD chips, wonder if they can be used at the telescope for astrophotography. The answer is a qualified "yes."

These are, in fact, real CCD cameras. They use detectors not too different from those found in astronomical cameras. But there are important differences as well. CCD cameras that are used for long-exposure deep sky photography are chilled to keep thermal noise in check. Filmless digital cameras are not cooled, as the longest exposure they'll ever have to make is a second or two. Even if there were a way to eliminate thermal noise *without* cooling, there is no provision on these cameras for making long exposures. The hard-wired computer programs that run consumer digital cameras don't offer an equivalent of a film camera's B exposure setting. Because of this inability to expose for very long, digital cameras are useless for deep sky work. But that doesn't mean they're *completely* useless in astronomy. They can make very nice pictures of the Moon and bright planets.

The main problem encountered by the digital camera owner who wants to take a picture of the Moon or Jupiter is that the lenses on all but the most expensive cameras is not removable. Because of this, there's no way to attach the camera to the SCT via eyepiece projection or prime focus adapters. Even if you *could* find a way to rig your camera to these accessories, it wouldn't work. You must take the lens off to shoot either prime focus or eyepiece projection images. But what you *can* do is use the same afocal method of imaging we mentioned in our discussion of film photography techniques.

Using your digital camera for afocal lunar and planetary photography is simplicity itself. All you do is point your CAT at the object of interest, aim the digital camera into the eyepiece and shoot. Focus by using the focus control on the scope, and control magnification either by eyepiece selection or by zooming the camera if it's equipped with a zoom lens. Try to get the lens as close to the eyepiece as possible without touching it to eliminate the vignetting "porthole" effect that will otherwise occur. If you wish, you can place your camera on a tripod set up behind the scope. This will help eliminate blurring caused by camera shake. Once you've got some nice shots, you can download them to the computer just as you would any other picture and use the software that came with the camera to enhance the images. Judicious use of cropping and brightness and contrast adjustments can produce pictures – especially of the lunar terminator – that are quite impressive.

Is there anything other than the Moon and bright planets within range of a digital camera? Double stars are a natural as are brighter open star clusters. Most cameras are capable of capturing stars down to magnitude 8 or 9. Can you hope to get any kind of results on galaxies, nebulae or globular clusters? Sadly, no. It's possible to record some trace of the very brightest objects, but a *barely* discernable wisp of the Orion Nebula is about the limit for these devices. In addition to the lack of long-exposure capability, the CCD chips in these camera are *very* small, just like those in astro CCD cameras. The resulting high magnification makes it even more difficult to photograph dim objects with your digital camera.

But being limited to the solar system is not a tragedy. The Sun, Moon, Jupiter, Venus, Mars and Saturn actually provide plenty of challenges for the astroimager. Once you gain a little experience in using your digital camera and telescope together, you'll find that you're producing images as good, if not better than anything that can be done with conventional film photography by the average amateur. But would you like to make pictures of the Sun's family that are *much better* than what can be done with film and SLR? If your family owns a camcorder you have a tool that will allow you to make startlingly detailed pictures of the Moon and planets.

Video Astronomy

Camcorders

Why is it so hard to make detailed photographs of the planets? Even large telescopes almost always produce pictures that at best show far less than what was visible to the eye through the same telescope. At worst, planetary photographs look like blurry messes. It is much easier to take a photo of a distant galaxy than it is to produce a detail-rich image of Jupiter. The reason for this is *seeing*. Atmospheric unsteadiness blurs out the tiny details that solar system enthusiasts prize. Sure, the atmosphere steadies down occasionally, but it is very difficult to take a photo at just the right moment to take advantage of these stable interludes. It is also a fact that the planets, when enlarged with eyepiece projection, are dim in an 8 or 10 inch telescope – dim enough to require exposures of perhaps several seconds. Often several seconds is long enough for periods of calm seeing to pass. Disturbed cells of air pass in front of the scope's view before the exposure ends, making the result bland and fuzzy. CCD cameras can help a lot with the latter problem. They are sensitive enough that they can make do with exposures of less than a second in planetary imaging. But it's still difficult to know exactly when to push the shutter release to take an image to capture a planet during times of best seeing.

I was contemplating these planetary imaging problems one summer evening a few years ago. Jupiter was riding high and was a mass of wondrous detail through the eyepiece of my 8 inch SCT. How I wished there were a way to *capture* some of that detail. I'd tried planetary photography numerous times over the past 25 years, and I'd gotten good results with the Moon, but the planets seemed hopeless. The best I'd ever done with Jupiter was a color photograph that showed a couple of low-contrast cloud belts – if you knew what to look for. The problem was always *seeing*, even in my area of the world, the southeastern US Gulf Coast, where the atmosphere is often stable. If only there were some way to know exactly when to take a picture of Jupiter. *Or a way to take many pictures in succession.*

What type of camera allows you to take many frames successively? A movie camera. Or a video camera. I'd been using an 8 millimeter consumer-grade camcorder for some time to take pictures of family events, but had never thought to try it at the telescope. Could a camcorder capture Jupiter through my CAT? I ran inside for my Sony 8 millimeter.

Most camcorders, like digital cameras, don't have removable lenses, so my first challenge was figuring out how to shoot through my SCT with a camcorder. I'd used the afocal method occasionally for film photography, so I knew this was the way to go. It was fairly easy to get Jupiter in the viewfinder of my camcorder by just pointing it into the 25 millimeter eyepiece in my star diagonal. What *wasn't* easy was keeping it there. My early 1990s model of camcorder was much heavier than today's miniaturized models. But with a little experimentation, I found that I was able to hold the lens nice and close to the eyepiece and keep Jupiter in view for extended periods of time.

In order to focus, I turned off my camera's auto focus system and sharpened up the image by using the SCT focus control. Initially, I was a bit worried about Jupiter being bright enough to show up on videotape. Even though my camcorder, like all vidcams sold today, used a CCD chip, I wasn't too sure how well exposed the planet would be. After all, a camcorder, like a movie camera, exposes each frame for only 1/30 of a second. But what I was seeing in my camera's viewfinder showed this wasn't a worry. At low power, Jupiter was actually *overexposed*. I had to bump up the power by using a shorter focal length eyepiece and increasing the zoom on my camera's lens before the exposure was acceptable. I could see that I was recording some details of the planet's atmosphere, but the camcorder's black and white viewfinder made it difficult to tell just how much. I continued to shoot for another 15 minutes, trying various eyepieces, filters, and zoom amounts on the camera in hopes of getting something.

Inside, I connected the camera to a television and hit play. I had absolutely no idea what to expect. What appeared on my screen amazed me! Not only was Jupiter huge on my screen – the magnification delivered by my imaging setup was equivalent to a magnification in the thousands – and it was also quite detailed. Not only could I easily make out several cloud bands, I could also see the somewhat pale Great Red Spot. The four Galilean moons showed up well in my lower-magnification, wider-angle shots. The only major problem I noted with my footage was that though I thought I was holding the camera steady, the high magnification was causing image shake, making Jupiter jump all over the screen and making me feel like I was coming down with a twinge of seasickness.

The next evening I was out with my camcorder again, but with a few refinements. Instead of hand-holding the camera and pointing it into the star diagonal, I set the camera up on a tripod stationed behind the telescope. I removed the star diagonal and inserted the eyepiece directly into the visual back. This made it easy to move the camera until its lens was almost touching the eyepiece. I found that I had to move the camera and tripod occasionally to recenter Jupiter as the telescope moved to track the planet across the sky, but that I could shoot for fairly long periods before this became necessary. On this second night out, I was rewarded with a Galilean satellite's shadow transit. On the finished tape, the shadow of the Moon on Jupiter's face showed up surprisingly well, appearing as a tiny hard black dot on the massive globe.

To say that I was pleased with my results would be an understatement. My videotape Jupiter was far better both in details and contrast than anything I'd ever done with an SLR. I also found, surprisingly enough, that I could actually see details on my tape that had escaped me visually. This was because of two factors. First, I could rewind the tape as often as I wished, observing Jupiter during moments of good seeing again and again. And there's also something to be said for making your observations of Jupiter in the comfort of your den instead of out in the bug-infested dark.

There was only one thing missing – stills. My tapes looked very nice, but I wanted some still pictures to show off to my friends and to post on the Internet. I knew that it was possible to obtain frame grabbers which allow you to capture single video frames and download them to the computer for processing and printing, but I had been under the impression that the equipment needed to do this was very expensive. But then a friend mentioned "Snappy." A Snappy is a small video-game-cartridge sized device that plugs into the printer port on the back of your IBM compatible computer. With the Snappy connected to the PC, you plug the video output from your camera or VCR into the Snappy. When you start the Snappy's software, you're presented with a small preview screen on your computer monitor. Press play on your camcorder and you'll see that, almost magically, your video appears in this window. When you see a frame you really like, you click your mouse and a single frame from your tape is captured. It can then be saved in a number of image formats for processing, just as you would with a real CCD image. With a little work with image processing, I was making stills that really looked like film photographs. Except that the lunar and planetary images on them were worlds better than I'd ever done with my film camera.

The above all sounds very interesting, but what kind of a camera is needed to pursue video astronomy? What kind of software is needed in order to process still images? And can you take pictures of anything besides Jupiter?

Camera choice is very straightforward. In fact, if you already own a camcorder give it a try. Chances are it will work or can be made to work. If you're in the market for a new camcorder, though, and would like to choose one that will work well in astronomy, there are some features you can look for. The most important characteristic for a prospective camcorder is sensitivity. How dim an image will the CCD chip of your vidcam register? The sensitivity of a camera is easy to determine since most manufacturers make this known in their advertising and camera specifications. Sensitivity is measured in units called lux. Smaller numbers are better, with a camera rated at 1 lux, for example, being much more sensitive than one rated at 3 lux. How low a lux rating do you need? In my opinion, in order to make the camera usable on dimmer subjects, you should insist on a value of about 1 lux.

Almost as important as the camera's sensitivity is its focusing mode. All consumer video camcorders on the market today are equipped with autofocus. This is fine for terrestrial subjects, but just gets in the way when you're using the camera at the scope. If at all possible, try to obtain a camcorder that offers the option of turning off autofocus, allowing you to focus manually. If this is not possible, look for one that at least offers *focus hold*. I've used these cameras for astro-imaging by pointing them at an object a few inches away, causing them to focus to their closest setting (which I've found is the best place to leave the lens during astrophotography), and pressing "hold." This makes the lens focus remain locked at this value and I focus, as always, with the telescope focuser.

What else does a good astrovideo camera need? A good tape format. Until recently, the consumer video buyer had three choices: VHS-C, 8 millimeter, and Hi-8. Which is best? This is somewhat subjective, but I'd rank their quality in this same order, with Hi-8 being by far the best. Things are changing rapidly now, with digital tape formats becoming more common and lower in price. I expect VHS and both 8 millimeter formats will disappear shortly – for camcorders, anyway. The coming of digital tape is all to the good for astrophotographers. Not only will the quality of the recordings be better, but it will become much easier to get your images into your computer. With digital tape, you should be able to eliminate the Snappy and other frame grabbing analog-to-digital gadgets.

Anything else? A few year ago I'd have advised you to try to locate a camera offering variable exposure settings. Unfortunately, the major camcorder manufacturers have almost abandoned such niceties for their consumer gear. The current trend is toward auto-everything camcorders with very little control being left for the user. Some cameras do still offer limited exposure control, but this is now often very much simplified, with a control labeled "darker or lighter" being more common that settings expressed in actual exposure values. I'm hanging onto my old camera, since it allows me to set the shutter speed from 1/30 second to 1/2,000 second.

If you're interested in producing stills from your videotapes you'll need some software to go with your Snappy. An image processing program will allow you to do things like improve the sharpness of your picture, adjust its brightness and contrast, and even stack several separate frames in order to produce one low-noise image. What's the best image manipulation program? In my opinion, Adobe's *Photoshop*. This program will allow you to do practically anything you can imagine to a photograph or any other kind of image. *Photoshop* is aimed at the professional publishing market, though, and is very expensive. A very reasonable alternative is the company's "junior" program, *Adobe Photo Deluxe*. Deluxe can do just about anything *Photoshop* can. The main difference is that the number of user adjustable controls and options has been reduced. This may actually be all to the good – most new users will find the full-blown *Photoshop* intimidating. Another imaging program favored by many astrovideo workers is the time-honored share-ware program *Paint Shop Pro*. This application really will do almost anything *Photoshop* will (though somewhat less easily and elegantly) and sells for far less.

What if you're not quite ready to invest in computers and frame grabbers, but would still like to make some still images from video? Fairly good results can be obtained by photographing the screen of your monitor or TV set with your SLR camera. Turn off all the lights in the room to prevent reflections off the television tube from appearing in your picture, frame your shot carefully, and expose for 1/30 second or slower to ensure that you don't capture TV scanning bars in your image.

What in the sky is available to a camcorder? Jupiter is a good subject for your vidcam, but it is not the only object your can capture easily. The Moon is an obvious target. Video can allow you to create amazingly good images of both individual craters and fairly wide-angle vistas. After a few months you'll be developing your own photographic lunar atlas. Video Moon pictures are so easy to make that you'll find that you can grab shots of most of the prominent craters at any given phase in an evening or two. Another great candidate for video filming is anything that moves. Lunar and solar eclipses, occultations, and shadow transits of Jupiter's satellites are examples of events that really take advantage of video's moving-picture nature.

As you keep moving out into the solar system with your camera, you'll come to the last detailed planet, Saturn. The sixth planet is a good camcorder destination, and you'll find that your 8 inch SCT can make nice images of the ringed world, even resolving Cassini's Division on a steady night. But it is at Saturn that the consumer video camcorder starts to show its limitations. Saturn is dim enough that it can be a problem for even a relatively sensitive 1 lux camcorder. The secret with this planet is to keep the image small enough to remain noise free. You'll find that as objects get dimmer and your camcorder strains to expose them properly, your images become more and more noisy. The only cure is a more sensitive camera.

Closed Circuit Surveillance Cameras

After using my camcorder on the Moon and planets for a few months, I was very satisfied by my results. It was obvious that video had great potential for use in astronomical imaging, but I was becoming frustrated by my camcorder. The major irritant was a lack of sensitivity. I just couldn't get a decent picture of Saturn. If I reduced magnification to the point where the planet was nice and bright and noise-free, it was so small that not much detail was visible. Better resolution would have been nice too. I was able to see the cloud bands of Jupiter, but I wanted to record more details in the atmosphere – the festoons and wisps that were easily visible in the eyepiece on good nights. A little more versatility would be welcome too. The afocal method of imaging worked, but being able to attach the camera directly to the telescope via my photo adapters would be highly desirable. Using a vidcam at prime focus would allow me to do wider angle shots of the Sun and Moon. Eyepiece projection could be used for high-power closeups of the planets and individual Moon craters. At this point I got lucky. I ran into an amateur astronomer who'd been doing video for a couple of years. He suggested that I look into small, inexpensive closed circuit (CCTV) cameras.

A little checking around turned up several distributors for video cameras of this type. A glance at their specifications showed that they might serve very well at the telescope. The first thing that got my attention was the sensitivity of these devices. Many are used in security applications where low light is the rule (they are often referred to as surveillance cameras). For this reason, these black and white cameras generally have lux ratings in the 0.03–0.01 range. This is obviously tremendously better than the average camcorders. CCTV/surveillance cameras also come equipped with removable lenses, usually using C-mount lens mounts. C-to-T adapters are easily available. A C-type T-ring will allow a surveillance camera to be used with all telescope photo adapters. I was also encouraged by the size and weight of the cameras. Since they are cameras only – they don't incorporate onboard tape decks – they can be very small and lightweight. Just a few inches long and less than a pound. This is smaller and lighter than many modern wide-field eyepieces. The prices for these cameras are also easily manageable. Black and white CCTV cameras can be found for less than US $100.

There is really only one area where surveillance cameras fail to improve upon your camcorder: convenience. Camcorder astronomy is easy. All you do is set up your scope as usual, point the camcorder at the eyepiece and shoot away. Using a CCTV camera means that in addition to the telescope and camera you'll also have to provide some means of recording your videos. The obvious solution is with your family VCR. You'll also need a way to view the images, since the little surveillance units don't have viewfinders of any kind. So plan on dragging a portable TV or monitor into the field. Cables will have to be run from camera to VCR and from VCR to TV, of course. Unless you're lucky enough to have a battery-powered TV and VCR available, you'll need a source of AC/mains power, too. Extension cords and power strips will need to be obtained and run. Quite a carload of gear. For this reason, I tend to do all my CCTV video work from my backyard. As long as I'm staying at home, setting up the equipment is not a huge chore, and I quickly found that my results justified the effort.

I was able to test my surveillance vidcam and my support equipment on Saturn shortly after obtaining my new camera. I won't pretend that setting up all the gear on the first evening was easy. Getting everything properly connected and powered up was frustrating. Even more frustrating was trying to put Saturn on the screen of the 12 inch television set I was using for a monitor. There is one other area in which CCTV cameras are inferior to camcorders: the size of their CCD chips. The detectors in inexpensive surveillance cameras are 1/3 inch in size – considerably smaller than those in camcorders and digital cameras. This means that it can be very hard to place the high-power image of a planet in the field. I found that the best way to do this was by making sure that my finder scope was very precisely aligned using a high power eyepiece in the main telescope. Even with the finder scope dead on, I had to do some searching with the telescope's slow-motion controls. But when I finally got the ringed wonder positioned on the TV screen, I knew that the expense and effort to upgrade to a surveillance cam had been worth it.

The images of Saturn on my screen and the resulting tapes were far better than anything I'd been able to achieve with my camcorder and 8 inch SCT (see Figure 10.5). The pictures I was getting were a lot better than some camcorder footage I'd made using a friend's *24 inch* Dobsonian! Not only was Cassini's Division visible, the Crepe Ring was also discernable, as were brightness variations across the rings, banding on the disk of the planet, and the planet's big moon, Titan.

Results with other objects were similarly improved. Detail on Jupiter was also much better. Smaller craters on the Moon could be blown up to reveal detail. Even Mars gave up some features on its small disk. I also found it possible to pull out fainter stars than was possible with the camcorder, meaning that many more double and multiple stars were within my range.

Should you try CCTV video-imaging? If you've exhausted what you can do with camcorders, an inexpensive surveillance cam may be the next step. Experimenting with these devices can also stand you in good stead if you intend to move on to a real CCD camera. Many of the challenges are similar. Finding, focusing, and image processing are all tasks that you can really get a taste of with a surveillance camera and a Snappy or other frame grabber.

Which surveillance camera? There are many surveillance CCTV cameras on the market, but you should choose one with good sensitivity (0.01–0.05 lux), and a removable C-mount lens. Some amateur astronomers are experimenting with color CCTV cameras, but these are invariably less sensitive and offer less resolution than their black and white counterparts.

Some potential video astronomers eyeing surveillance camera setups wonder if there's a way to avoid carrying all that gear into the field. Wouldn't it be possible to run the camera's video directly to a laptop computer via a Snappy frame grabber, eliminating the need for a monitor, a VCR and AC power (most CCTV cameras can be powered by any 12 volt DC source)?

Figure 10.5. Video image of Saturn.

This could be done, and some astrovideographers (a word those of us working in astronomy video have coined) are doing it. But in my opinion it makes you forego video's prime advantage. When you're shooting video, you are taking 30 frames every second. The resulting huge amount of still pictures you'll have available on tape means you're just about certain to get some frames during moments of excellent seeing. If you go the Snappy/laptop computer route, you'll be back to trying to guess when the seeing is stable before snapping a frame. Inexpensive frame grabbers like the Snappy, you see, are very limited in their ability to capture more than one single frame at a time.

Computers and CATs

CATs and computers go together perfectly. The versatile nature of our telescopes means they're easily adaptable to computerization. If you've purchased a goto scope like a Meade LX-200 or a Celestron Ultima 2000 or if you've added a set of digital setting circles to your older or non-goto scope, you've taken the first step toward allowing a computer to help you with observing tasks. But gaining a real foothold on the leading edge technology-wise means using a genuine personal computer with your scope.

A Laptop at the Telescope

If you're fortunate enough to own a laptop computer, either a PC or a Macintosh, you already own a very important part of a computerized telescope system. With your laptop connected to a port on either your goto scope or your digital setting circle computer, you'll be presented with a star map on your computer display. Somewhere on this display you'll see a cross-hair reticle. This will indicate the actual position of your telescope – where it is pointed. If you've got a goto scope, you can center the computer display on an object in the program's database, click on this target with your mouse and have your scope slew to the target! If you're using digital setting circles rather than a goto scope you'll have to move the scope by hand. On the computer the cross-hairs will move across the screen's star display as you move the telescope. Once the reticle is centered over the object of interest on the PC screen, it will also be in the field of view of your telescope.

What's the advantage of using a PC in conjunction with a goto scope or a DSC computer? One *huge* advantage is the size of the databases offered by many astronomy programs. You may think that the number of objects stored in the memory of your LX-200 or DSC is big, but this pales in comparison to what's on the CD ROMs of most deep sky computer applications. Well over a hundred thousand objects is now very common. Most of these will be dim galaxies from the professional catalogs like the PGC (*Principal Catalog of Galaxies*) or UGC (*Uppsala Galaxy Catalog*), but having just about any object in the sky available can be handy for astrophotographers and CCD imagers, and it will certainly impress your friends. But the biggest advantage to using the computer is that it makes things *easier* in the field.

The LED or LCD displays on goto scopes and DSCs get the job done, there's no doubt about that. But it is far easier to select objects by looking at a graphic representation of the sky rather than a set of numbers on a small DSC display. A PC display also shows you what's available in the same area as your initial target. You can tell which other interesting objects are nearby without having to resort to flipping through a star atlas. If you're using digital setting circles, the PC is an incredible improvement. Normally you have to watch the small DSC LED display as you move the scope until either the right numbers are displayed or until small arrows or other indicators that show you which way to move your scope indicate you're on the right spot. It's just so much nicer to be able to watch how a cross-hair moves across the computer star field and to stop moving when you're pointing in just the right direction.

What Do You Need?

If you're convinced you're ready to become a part of the computer-telescope revolution, you'll need more gear. What's required? The first thing you need is hardware on the telescope end to allow you to connect PC to SCT. If you use a goto telescope all that's required is an appropriate serial (RS-232) cable to hook scope to laptop. These can be purchased from astronomy vendors. If you own a manual CAT you will also have to add a set of digital setting circles. The brand is unimportant. What *is* important is that the DSC computer possesses an RS-232 serial port that will allow you to connect it to the computer's serial port.

If you already own a set of DSCs that don't feature RS-232, all is not lost. *Computer–telescope interface boxes*, also called B boxes, are available for use with your current optical encoders (the encoders are the devices attached to the RA and declination axes of your scope that tell the DSC computer where your scope is pointed). Unplug your DSC computer, plug the B box in its place and connect the box to your serial cable. Since B boxes don't have displays or much in the way of electronics, they are relatively inexpensive. If you intend to always use your scope with a PC, all you may need is a B box. For those who don't already own DSCs, B boxes can be purchased with encoder sets appropriate for your particular telescope.

No matter what kind of telescope setup you have, you'll need a PC of some kind too. As is the case with CCD users, I don't recommend that you use a desktop in the field. Using an AC operated device (especially a high-voltage monitor) is a recipe for disaster in the damp and dewy outdoors. Not only is a laptop safer, it will free you from the necessity of being near AC lines. A laptop can be a considerable investment if you don't already own one. Luckily, though, unlike CCD processing software, most deep sky and planetarium programs do not require the latest and the greatest computers to run well. An earlier, slower-speed Pentium or even a 486 processor is all that's needed for many applications. These computers can be found for surprisingly low prices on the used market. If possible, try to find a laptop that can be operated off external DC as well as internal batteries. Running your computer off the big 12 volt battery you use for the scope drive or dew zappers means you won't have to worry about the computer's puny onboard batteries giving out in the midst of an observing run.

The final element needed to computerize the CAT is software. The choice of program is very important, and will in large part determine how pleasant to use and efficient your setup will be. Though amateur astronomy is a small field, you'll be amazed at how many software packages are available. Some can be eliminated right away. Unless you'll be content just using your laptop as a computerized star atlas, you'll need a program that is designed to be interfaced with your equipment. It must be able to work with your particular setup: LX-200, Ultima 2000, Nexstar 5 or ETX. If you're a digital setting circle user, it must be capable of interfacing with your brand DSC computer or B box.

In addition to being able to operate with your scope or DSC computer, you'll want a program designed to be used in the field by a *working observer*. Astronomy programs fall into two categories at present: planetariums and deep sky programs. Planetariums are "pretty" programs. They attempt to deliver photo-realistic representations of the sky. They may also include many multimedia frills – movies, sounds and computer aided instruction in astronomy. Many of them can be used with a telescope or DSC systems, but you probably won't want to. The displays that are so pretty in the house become confusing and irritating on the observing field. The full color screens ruin your night vision, the photo-realistic representations of deep sky objects make them hard to see, and the bleeps, bloops and music irritate your fellow observers.

What you want is what computer-packing amateurs call a *deep sky program* (though these applications can also be used for solar system tasks). These software packages are not nearly as beautiful to look at as the planetariums, but they are designed for use by *astronomers*. They offer many advantages over planetarium software. One plus is larger databases (in addition to hundreds of thousands of deep sky objects, most include the *Hubble Guide Star Catalog* of millions of stars). They also tend to offer control facilities for *many* different telescope and DSC systems. You needn't worry about ruining your hard-won dark adaptation either, since many of these programs can turn the computer display night-vision red. Most of these programs can also, when needed, print maps with quality comparable to that of the very best printed star atlases.

Which program? Astronomy is a growing hobby, and new releases appear all the time, but these are some of the programs I've used and which I feel are worthy of your consideration.

Megastar

This long-time favorite was written by US amateur astronomer and software artist Emil Bonano. *Megastar* is the program I've used more than any other over the last five years. Its database is second to none, and includes a selection of objects not available in other programs, the MAC catalog. This is legendary deep sky observer Larry Mitchell's *Mitchell Anonymous Catalog* of thousands of faint, formerly uncataloged "anonymous" galaxies. *Megastar* does a fine job of telescope/DSC interfacing too. The only place where the program falls down, in fact, is in showing you what the sky looks like for a given time and date. *Megastar* was designed from the first as a computerized star atlas. It is as far removed from a computer planetarium as you can get. Consequently, it barely shows you where the horizon is (the horizon is depicted as a single line). This makes it a bit difficult to orient yourself in regards to where things are in the real sky. Despite this one lack, *Megastar* remains my favorite.

Deepsky 2000

Steve Tuma's DS2000 is a fairly recent addition to the deep sky program lineup. It is also quite different from the other programs in this category. It uses a different "metaphor" – a different user interface – than the others. When it's started, *Deepsky 2000* presents the user with a spreadsheet screen rather than a computerized depiction of the sky. This is a different but surprisingly effective way to do things. *Deepsky 2000*, once you get the hang of it, is incredibly complex and capable and can do anything the others can and more. Just because it doesn't draw a star chart as soon as its started doesn't mean it can't plot excellent maps, either. It is capable of producing and printing atlas-quality charts. Due to its database orientation it really excels in generating lists and logs of deep sky objects and performing database searches. This CD ROM based program is also one of the most attractively priced astronomy applications on the market.

Skymap

Skymap, now in version 6.0, by the UK's Chris Mariott is nearly as good as *Megastar* at depicting the deep sky. And it is undeniably better at orienting you with the real sky – horizon, zenith, etc. There's little bad to say about *Skymap*, it's a mature and capable program capable of producing charts second to none. It is also very well-suited for telescope and DSC interfacing. One real advantage of *Skymap* is its author's responsiveness to user needs. Chris is often found online on popular Internet astronomy forums like the newsgroup sci.astro.amateur, and is more than willing to answer questions and listen to suggestions. If you're interested in the solar system, *Skymap* is without doubt your best choice. In tests, its depiction of solar system phenomena was found to be decidedly more accurate than those of the many other astronomy programs evaluated.

The Sky

Do you want an astronomy program that can do almost *anything*? *The Sky* doesn't stop at printing charts or controlling an LX-200. It can also interface with optional program modules that operate CCD cameras and control telescopes over the Internet. *The Sky* also contains a database quite comparable to *Megastar* and *Skymap*, too. The only drawback is its price. The program is sold in various levels, and the most advanced, Level 4, which is what you'll want, is over twice as expensive as *Megastar* and *Skymap*. Don't do Windows? Macintosh fans will be pleased to learn that *The Sky*'s publisher, Software Bisque, has recently ported this fine program to their favorite computer.

Guide

Guide is the dark horse. It offers all the features of the other programs and some that many of them eschew, like comparatively realistic animation and depiction of solar system objects. Database-wise, this program is right up there with other deep sky powerhouses. It hasn't yet caught on with users like the three previous programs. But it should and still may. That Guide must have something going for it becomes obvious when you listen to the comments of its users. They are definite and unashamed in their opinion that Guide is "the best."

Voyager

Lots of people like the Apple Macintosh. But "lots" translates into "relatively few potential users" when you're talking about a specialized area of software like astronomy programs. Voyager is the only readily available deep sky program for the Macintosh other than the recently issued *The Sky*. Voyager was actually the *only* deep sky program available for the Apple for many years. It is then, as you'd expect, a very full-featured and mature program. It is able to display deep sky objects and interface with telescopes with the best of 'em.

In the Field

With your system of telescope, computer and software ready, you're prepared for a night of computer fun in deep space, right? Almost, but not quite. You'll want to make provisions for protecting your laptop computer from dew and keeping it working in cold temperatures. A cover to keep dew off your PC can be as simple as a plastic sheet you place over it when it's not in use (leave a few air gaps between sheet and computer to allow some of the heat the laptop generates to escape) or as complicated as a ventilated box-style cover with a clear Plexiglas top to allow you to view the screen while the PC is shielded from dew.

Laptop computers run into two electronics problems in the cold: loss of battery power and dimming of the display. Power problems can be eliminated if you run the computer, as advised, off a large 12 volt battery. Your display will probably be OK if you keep the computer in a dew proof enclosure of some kind. The computer's self-generated heat will keep the cold sensitive LCD display working.

Before connecting cables between scope and computer, please be sure everything is powered down. You can then boot everything up and enjoy a night of efficient, futuristic observing. The first evening or two under the stars may be a bit frustrating. You'll have considerable extra gear to mess with, and getting everything operating will be a real chore to begin with. But before long initializing the computers will be as normal and easy as pointing your wedge toward north. Just remember to look up at the beautiful night sky once in a while!

If you've come this far in the book, you should now be ready to enjoy all of the wonders your universe can display. You are now skilled in the setup and operation of your SCT or other CAT. You know how to polar align and how to take photos with your telescope. You can even connect it to a computer system to help you take CCD images or aid you in hunting down the very faintest objects eye or camera can detect. But what do you do with all this stuff? What do you look *for*? What should your *goals* be? *How do you keep going? How do you keep your interest alive, voyaging into the sky night after night and year after year?*

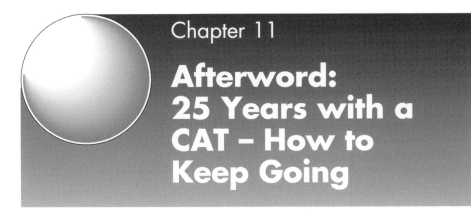

Chapter 11

Afterword: 25 Years with a CAT – How to Keep Going

Earlier in this book, I mentioned the "CAT in the closet" syndrome. Some people become enthusiastic about astronomy, buy an expensive telescope, use it a few times, and deposit it in a closet. Other astronomers get hold of a scope, aim at the stars and keep going year after year. I've been observing for 35 years now, 25 of those with our favorite telescope, the SCT. How did I manage to do it? How did I remain interested in this hobby month after month and observing season after observing season for over three decades?

One of my secrets is that I've *never* looked upon amateur astronomy as a mere *hobby*. It's been much more to me than that – I've always approached it as my alternate vocation. Even though I am now employed as an astronomy educator, I don't take this wonderful science any more seriously than I did when I was just an 11 year old with a flimsy 3 inch Japanese telescope. Even then astronomy felt like a way of life rather than a simple pastime. But other than making the practice of astronomy an important priority in your life, what else can you do to keep your amateur career alive over the coming decades?

Goals

The surest way to keep your astronomical life interesting is to set and work at goals. The prime reason that CATs hit the closet is that the owners have decided that they've "seen everything there is to see." Questioning these individuals will usually reveal, though, that they've hardly seen *anything*. Most have barely scratched the surface of the thousands of objects available to an 8 inch telescope. "Everything" usually turns out to actually be only the Moon, Jupiter, Saturn, Mars and a few of the very brightest Messier objects.

Why would a person think he or she has seen everything when very little has been observed? Usually because of a lack of concrete goals. These amateurs drag the scope out into the backyard, look at any bright planets that may be visible, point at a Messier or two, stand around for a few minutes and pack it in. If you go outside without solid ideas about what you're going to look at, you won't see much of anything at all. Before beginning an observing run, develop a list of objects you'd like to observe. For best results, concentrate on one constellation per evening. In addition to bright and well-loved objects, round out your plan with some objects you've never seen before. Finish your list with some objects you judge might be hard or even *impossible* from your location with your telescope. Your planning tools may be as simple as a star atlas and a loose leaf notebook or as high-tech as a computerized charting and planning program like *Deepsky 2000*.

Under the stars, *stick to your list*, hunting one object after another. I find it useful to save the spectacular Messier objects for the end of the evening as a treat. After an evening of hunting dim galaxies, a bright Messier globular cluster looks all the better. The key to observing success is to really push yourself. Don't give up on supposedly impossible objects. Push your observing skills to the limit. You'll find at the end of the evening that you have a real feeling of accomplishment – whether you've managed to view every object on your list or not. Rather than becoming bored and bringing the telescope inside after an hour or two, you'll start complaining that nights are too *short* and find yourself greeting the dawn more and more often. Short weeknight runs will turn into all-night marathons. You'll look and feel like a zombie at work or school the next day, but the fire in your belly will definitely be lit – you won't be able to get *enough* of the night sky.

Awards

Some amateurs like pursuing more structured goals. They want something more definite to shoot for than just working through a self-made list of deep sky objects. Awards for observing various groups of objects are available from many astronomical organizations around the world, most notably from the Astronomical League in the United States. The League offers a variety of Observing Clubs, but the place where most people start is with The Messier Club certificate. Two classes of this much-coveted award are offered, "standard" and "honorary." The standard certificate is given to any amateur who successfully logs observations of 70 of the objects in Charles Messier's famous catalogue. The honorary award is reserved for those deep sky workers who manage to find all 110 objects.

Astronomy Clubs

Along with having observing goals, the surest way to keep your amateur astronomy pursuit on the straight and narrow is to join an astronomy club. This is something I preach continuously. There is a joy in being alone under the stars with your telescope, but you'll find observing with groups of like-minded people is a welcome change. You'll also soon realize that the enthusiasm of your fellow club members reinforces your own.

Scientific Contribution

Though relatively few amateurs wind up making earth-shattering discoveries in astronomy, this is an area many like to at least dabble in, often by hunting for comets. And for a lucky and dedicated few this pays off – a new comet bearing the discoverer's name lights up the heavens. But there is a way that any interested amateur can be *assured* of making real contributions to the science: by observing variable stars. For decades

AAVSO, the American Association of Variable Star Observers, has organized amateurs to provide a stream of data to professionals. Observing single stars and judging their magnitudes is not everyone's cup of tea, but who knows? You may find it as addictive as have several generations of AAVSO variable star observers.

Buy a Different Telescope

It's a good thing CATs are (relatively) cheap, because you may find a new one can easily make observing fun again. Especially if the new telescope is ultra-portable. The average 8 inch SCT is supremely *transportable*, but it is big enough that it's not something you'll be able to stuff under the airline seat ahead of you or even throw in the trunk with the suitcases when you're going on vacation. If you've got a larger CAT, just setting up out back is a somewhat monumental exercise. Sure, a Nexstar 5 or a 3.5 inch ETX or Questar can't compete with your "big gun" for sheer light-gathering power, but these little scopes are so easy to set up you may find you're using them a lot for quick weeknight sessions. A yearly ho-hum visit to the mountains or seashore may become really special if you can take a capable telescope with you. And these little telescopes *are* very capable. A computerized Nexstar or ETX may enable you to actually see more during a short run, despite the small aperture, than you normally do with your 8 or 10 inch SCT.

New Accessories

Sometimes you don't need to buy a new telescope to refresh your astronomical soul. A new eyepiece or other accessory may be all it takes. A new set of eyepieces, especially, can make your old warhorse orange-tube C8 into a brand new deep space machine. If your eyepieces are more than five years old or are of simple designs like Kellners and Orthoscopics, investing in one of the new design ultra-wide-field eyepieces can be a remarkable experience. Going from an apparent field of 40–50 degrees to 65–85 degrees really is like turning on the lights in a dark room!

New Pursuits

Anything can get boring after a while – almost anything, anyway. If you're feeling astronomically turned off, it may be time to change your astro lifestyle. If you're a dyed-in-the-wool visual worker, maybe trying the art of astrophotography or CCD imaging will provide the ticket out of dullsville. Planetary fanatic? Dust off that star atlas and start chasing faint fuzzies, deep sky objects.

Enter the Larger World of Astronomy

Joining your local astronomy club is a good way to keep going, but the local scene is not all there is. Throughout the US and Europe star parties are becoming increasingly popular. A star party happens when amateurs from a region, a country, or the entire world assemble at a remote and dark location for a few days to a couple of weeks of observing and camaraderie. These events owe much of their popularity to the growth of light pollution. Beginning in the 1970s, the average amateur found it more and more necessary to travel to dark locations to do real deep sky observing. Those areas blessed with really dark sites began to attract amateurs from far afield. What began as informal gatherings became more organized, offering convention type activities as well as observing.

The Internet

Don't feel like you can travel thousands of miles to a star party, but would still like to associate with your fellow amateurs throughout the world? Try the Internet. The Usenet Newsgroup sci.astro.amateur (s.a.a.) is online 24 hours a day and is inhabited by some of the most knowledgeable observers you'll find anywhere. While the level of expertise on s.a.a. is almost overwhelming, beginners are not only welcome, they are *encouraged* to participate. S.a.a. like other newsgroups is a "super bulletin board." You post a comment or question and other s.a.a. users respond to it. If you have a question about observing, telescopes or the science of astronomy, this is the place to have it answered.

Similar to s.a.a. but smaller and a little less daunting are the astronomy mailing lists. Mailing lists work very simply. You send an email to the address of the list server just like you'd send any email. The listserver computer then resends this message to everyone subscribing to the mailing list. The email you sent is received by the entire group. Many people like mailing lists because they are moderated. The person who started the mailing list usually serves as this moderator, enforcing list rules to keep things running harmoniously. "Flame-wars" and other insult contests are generally not tolerated on the mailing lists, making them a bit friendlier than the free wheeling s.a.a. Your author currently runs a mailing list specifically for CAT users, "sct-user." Despite the title, this list also welcomes owners of MCTs and all other catadiopric scopes.

Keep your Balance

Astronomy is the Queen of the Sciences. What she offers is both aesthetic and intellectual beauty. Unfortunately for many amateurs, though, this giver of beauty is a source of friction with family members, especially non-astronomer spouses. What do you do if your husband or wife simply can't understand *why* you want to spend "all that money" on telescopes and accessories only to stand out in the middle of the dark yard alone all night? The advice usually given to people in these circumstances is to keep things balanced. Don't neglect your family to pursue your observing programs. This is good advice, and I recommend it.

But what do you do about the husband or wife who becomes seriously upset even though you *don't* neglect your family? When all you want to do is observe for a few hours once a week? You should, of course, make clear that astronomy is *what you do*. It's what makes you *you*, and that it is *very* important to you. Sometimes showing is better than telling, though. Involve your Significant Other. Instead of packing up the scope, jumping in the car and leaving your husband or wife in the dust as you head to the club dark site, why not invite him or her along? Pack a couple of lawn chairs. Maybe even a bottle of wine. Set up the scope as normal, but also spend some time with your beloved (husband or wife, I mean, not the telescope!) just looking at the constellations, pointing out the naked eye and binocular wonders and telling their stories. When you run across a particularly interesting object, center it in the scope for your companion. And don't hog the eyepiece. In this way, the scope becomes a friendly object rather than a rival.

As dawn breaks, open that bottle and toast the star dome and your love! Don't patronize your spouse or talk down to him or her either. Just do your best to try to explain why you're hooked on the sky. I have no doubt that in a setting like this a little bit of the enthusiasm will rub off! Maybe not enough to make your mate want to observe the sky every night like you do, but enough to give some understanding of your *magnificent obsession*. Don't let this be a one time good deal, either. Involve your mate in your astronomical endeavors as often as he/she desires. Many star parties include activities for non-astronomer spouses and children, so one of these events can serve as an interesting alternative family vacation. And who knows? You may turn around one day to find you've created a monster. Your spouse has been bitten by the bug and has now laid claim to the "family" CAT!

And Keep the Stars in your Eyes

All of these ideas are proven ways to keep your astronomical interest alive, to make your relationship with your CAT a long and happy one. There is one final bit of advice I can give, though, one thing that will *really* see you through over the long run: *keep the stars in your eyes*. Try at least occasionally to remember *why you came here*. The reason most of us entered astronomy and started gawking at those pretty Schmidt–Cassegrain advertisements was because of our basic *sense of wonder*. Because every time we looked up at the night sky, we were filled with an almost mystic curiosity about the Great Out There! If you feel a little burned-out, try to recall these feelings of awe about the mysteries of our majestic universe! *Remember the first time you looked at a globular cluster and wondered if anybody were looking back?* And that is the only real secret to becoming an astronomy old-timer. Forget the minutiae of eyepiece designs, PEC recordings and CCD chip sizes once in a while and *wonder* again. If you can do that, your journey of discovery with your beloved CAT will be a long and fruitful one.

"May the stars light the end of your road!"

Index